EARTH ANGLES

PRECALCULUS MATHEMATICS WITH APPLICATIONS TO ENVIRONMENTAL ISSUES

PREVIEW EDITION

NANCY ZUMOFF • CHRISTOPHER SCHAUFELE

KENNESAW STATE UNIVERSITY

WITH CONTRIBUTIONS FROM MARLENE SIMMS AND PETER WALTHER

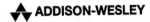 ADDISON-WESLEY

An imprint of Addison Wesley Longman, Inc.

Reading, Massachusetts • Menlo Park, California • New York • Harlow, England
Don Mills, Ontario • Sydney • Mexico City • Madrid • Amsterdam

This work is supported by U.S. Department of Education, FIPSE, grant #P11A30555, and National Science Foundation grant #DUE-9354647

Reproduced by Addison-Wesley Publishing Company Inc. from camera-ready copy supplied by the author.

ISBN   0-321-02858-9

4 5 6 7 8 9 10 CRC 009998

CONTENTS

CHAPTER ONE
FUNCTIONS

Introduction

In this chapter we will look at functions in different ways. We will study functions formally, their domains and ranges, but will also concentrate on interpreting graphs and understanding the meaning of functions used in other disciplines and those found in newspapers, magazines and periodicals.

1.1   Examples of Functions

A *relation* assigns to each number in a particular set (called the domain) at least one number from another set (called the range). A collection (set) of data describing a real phenomenon defines a *function* when the relationship assigns a single number in the range. For example, in the table below years are listed in the first column and the second column lists the peak flow of the Columbia River at the Dalles, Oregon, in thousands of cubic feet per second (cfs):

Year	Cubic feet of Water/sec (thousands)
1950	744
1960	470
1970	425
1980	345
1990	391

This describes a function. The domain is the set of years

{1950, 1960, 1970, 1980, 1990}

and the range is the set

{744, 470, 425, 345, 391}.

For each of the indicated years the function gives the peak cubic feet per second of flow in the Columbia River, i.e., to each year a single value is assigned for the volume.

Another way to look at a function is graphically. For example, Figure 1.1 below shows the consumption of bottled water annually (in billions of gallons) from 1988 through 1995.

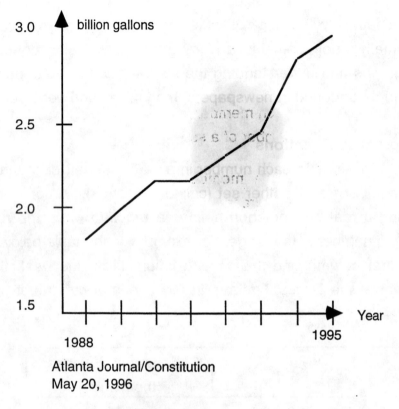

Atlanta Journal/Constitution
May 20, 1996

Figure 1.1

This is also a function. For any year between 1988 and 1995 there is a unique number (a single value) which indicates the consumption of bottled water for that year. For example, in 1992 the annual consumption of bottled water was approximately 2.3 billion gallons.

Another way of defining a function is with a mathematical expression, i.e., a formula relating two quantities. For example, the equation

W = 26P

describes the total daily water consumption W (in cubic feet) in terms of US population P. If the population is 2.52 million then the daily consumption is 65.52 million cubic feet.

1.2   Definitions and Notation

A function is a mathematical way of expressing the dependence of one quantity on another quantity. Functions occur in many real world situations: the area of a circle is a function of the length of its diameter; water consumption in a certain town is a function of the population of that town; the cost of mailing a package is a function of its weight.

We formally define a function as follows:

a *function* assigns to each member of a specified set, called the *domain, exactly* one member of a second set, called the *range.*

When a variable represents a member of the domain, it is called the *independent variable*; when it represents a member of the range, it is called the *dependent variable.* We sometimes refer to the independent variable as the "input" and to the dependent variable as the "output". Note that the above definition does not preclude more than one value in the domain from having the same corresponding value in the range.   For example the maximum temperatures on consecutive days of a week are given in the table below.

Day	Sun.	Mon.	Tue.	Wed.	Thur.	Fri.	Sat.
Temperature	42º	44º	45º	48º	44º	44º	41º

The domain is the day of the week and the range is the corresponding temperature. Here three values in the domain share the same value in the range. This is still a function under the definition given above, because for each value in the domain there is one and only one value in the range.

Let's look at another example. The following table gives US. population in millions in the indicated year:

Year	1960	1970	1980	1990
U.S. Population (millions)	181	205	228	250

Source: Statistical Abstracts of the United States, 1993.

The independent variable represents the year and the dependent variable represents the U. S. population. Since the table gives a single population value for each year, it represents a function. The domain is the set

{1960, 1970, 1980, 1990}

and the range is the set

{181, 205, 228, 250}.

Usually functions are named with letters such as f, g, or h. Name the population function by the letter f. Since population P is a function of year t we write P = f(t). This is called *functional notation*. The function f assigns to any year t in the domain the population P in year t. Read this notation as "P equals f of t." Here P is the dependent variable and t is the independent variable. Looking back at our first example, with regard to the Columbia River, we might use F = h(t) where function h assigns to any year t in the domain the peak flow F in the range. Thus

$h(1960) = 470 \times 10^3$ cfs.

* * * *

Next consider the following table.

x	1	2	3	5	2
y	0	3	1	6	8

The independent variable, or input, is x and the dependent variable, or output, is y. To decide if y is a function of x you must decide whether each different input value x produces exactly one output value y. Notice that when x = 2, y can be either 3 or 8. This means that the table does not represent a function. We say that this table represents a relation, but not a function.

The graph in Figure 1.2 shows the demand for water in Atlanta, Georgia, (vertical axis) during July from 1990 through 1995 (horizontal axis).

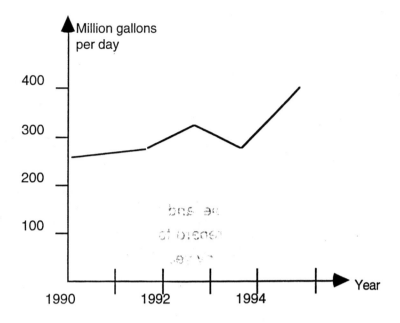

Water Demand in millions of gallons per day

Figure 1.2

We can see by looking at the graph that, for each year indicated on the horizontal axis, there is a unique corresponding value representing the demand for that year. Thus, the graph represents a function. Notice that a vertical line drawn through any value on the horizontal axis intersects the graph at most once.

A quick way to determine whether or not a graph represents a function is to use the *vertical line test*.

Vertical Line Test: If no vertical line intersects a graph more than once, then the graph represents a function.

Example 1.1

Which of the graphs shown in Figure 1.3 is the graph of a function?

(a)   (b)   (c)

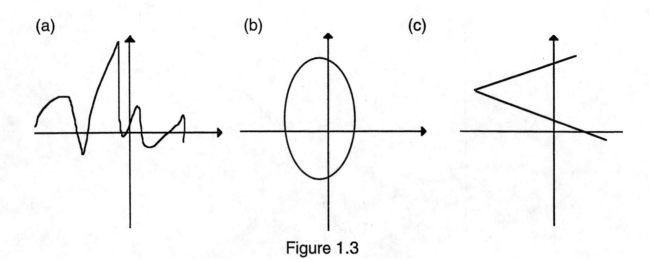

Figure 1.3

Solution

Only graph (a) is the graph of a function since no vertical line intersects the graph in more than one point. For (b) and (c) there are vertical lines that intersect the graphs in more than one point.

Class Work

1.   Use functional notation to express the statement "the dependent variable p is related to the independent variable q by the function F".

2.   Which of the following graphs represents a function?   Explain you answer.

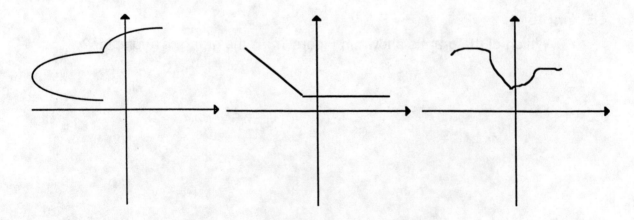

3.   From the world around you, select an activity or real phenomenon that can be modeled by a function. Represent this function using functional notation, a graph, and a table.

* * * *

As another example of a function, let's look at the relationship between temperature in degrees Fahrenheit and temperature in degrees Celsius. If we know the temperature in degrees Fahrenheit F we can determine the temperature in degrees Celsius C by using the formula

$$C = \frac{5}{9}(F - 32) \ .$$

If we name the function S and use functional notation,

$$C = S(F),$$

this expresses the fact that we are thinking of C as a function of F. For example, if $F = 50^0$,

$$C = S(50) = \frac{5}{9}(50 - 32) = 10$$

means that a temperature of 50º Fahrenheit corresponds to a temperature of 10º Celsius. Similarly, if $F = 94^0$,

$$C = S(94) = \frac{5}{9}(94 - 32) = 34.4 \text{ (rounded to one place)}$$.

means that a temperature of 94º Fahrenheit corresponds to a temperature of 34.4º Celsius.   In the above procedure, we have *evaluated* the function at 50 and 94, respectively. The graph of the function is shown in Figure 1.4.

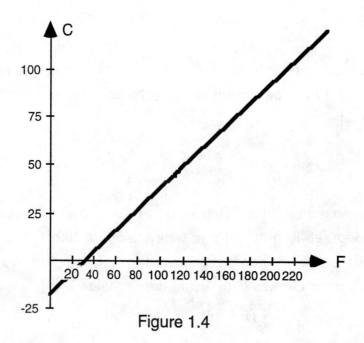

Figure 1.4

In the above example of $C = S(F)$ we are expressing C in terms of F; that is, $S(F) = \dfrac{5}{9}(F - 32)$. What if we wanted the function which expresses F in terms of C? We will call this function T and write $F = T(C)$, where F is in degrees Fahrenheit and C is in degrees Celsius. We start with the formula $C = \dfrac{5}{9}(F - 32)$ and solve it for F in the following steps:

1. $\dfrac{9}{5}C = \dfrac{9}{5} * \dfrac{5}{9}(F - 32)$ Multiply both sides of the equation by $\dfrac{9}{5}$.

2. $\dfrac{9}{5}C = F - 32$ Simply the right side.

3. $\dfrac{9}{5}C + 32 = F - 32 + 32$ Add 32 to both sides of the equation.

4. $F = \dfrac{9}{5}C + 32$ Simplify and rearrange the equation.

5. $F = T(C)$ Here we substitute $T(C)$ for $\dfrac{9}{5}C + 32$

which is what we wanted.

8

You can evaluate functions with a calculator or with computer software, but each does it a little differently. See your manual for instructions.

Class Work

Evaluate each function at the given value. Use your calculator or computer.

3.   Given: $l(x) = 3.21x^2 - 2.56x + 1$; find $l.(-1.05)$.

4.   Given: $s(t) = \dfrac{3t}{t^2 + 7}$; find $s(5.6)$.

5.   Given: $B(r) = (2r + 3)^4$; find $B(-0.2)$.

$$\ast\ \ast\ \ast\ \ast$$

1.3   Domain and Range

In the following subsections, we study set where a function is defined and where its values lie.

1.3.1   Domain

If the domain of a function is not specified, we usually take it to be the set of all real numbers for which the function is defined. For example, the function
$$g(x) = \frac{2}{x-3}$$
is not defined when x = 3 since division by 0 is not allowed. The only problem in evaluating g(x) occurs when the denominator equals 0, that is, when x = 3, so the domain of g is the set of all real numbers except 3. In set notation we write
$$\{x | x \neq 3\},$$
or in interval notation
$$(-\infty, 3) \cup (3, \infty).$$
(We will express the domain of a function using set notation, interval notation, or words, whichever is most convenient.)

What about the domain of $f(x) = x^3 - 3x + 1$? Since any real number can be substituted for x, its domain consists of all real numbers.

A different kind of situation is presented by the function
$$f(x) = \sqrt{x}.$$

9

No negative numbers can be substituted for x because you can't take the square root of a negative number and get a real number. But anything else is OK to substitute for x, so the domain of h is the set of all non-negative real numbers,

$$\{x \mid x \geq 0\},$$

or, in interval notation, $[0, \infty)$.

There are also practical considerations to take into account when the function describes a real situation. For example, the function

$$D(s) = 0.05557s^2 - 0.00143s$$

is defined for any real number s. However, this function can be used to model the stopping distance of an automobile when the speed is *s* mph. It would make no sense to substitute a negative number in place of *s* in that situation, so the domain of the function would be $\{s \mid s \geq 0\}$.

Example 1.2

Determine the domain of $f(x) = 3x + 5$.

Solution

Since any number can be substituted for x, the domain consists of all real numbers, or in interval notation, $(-\infty, \infty)$.

* * * *

Example 1.3

Determine the domain of $g(x) = \dfrac{x+2}{x^2-1}$.

Solution

Since division by 0 is undefined, we must find what value(s) of x will make the denominator equal to zero. We can do this by solving the equation x2 - 1 = 0.

$$x^2 - 1 = 0$$
$$x^2 = 1$$

$$x = \pm 1$$

Thus the domain of the function g is $\{x \mid x \neq \pm 1\}$.

* * * *

Example 1.4

Determine the domain of

J(p) = the unemployment rate (percentage of people unemployed)

when the gross domestic product is p.

Solution

The independent variable p denotes the gross domestic product, the monetary value of all goods and services produced in the country. Thus $p \geq 0$ and the domain of the function J consists of all non-negative real numbers, in set notation, $\{p \mid p \geq 0\}$.

* * * *

Class Work

Determine the domain of each function.

6. $f(x) = 3x - 1$

7. $p(w) = \dfrac{w}{w^2 - w - 2}$

8. $g(t) = \sqrt{4 - t}$

9. The function, $M(T) = .03(T - 32)$, describes the number of inches of snow that melts each day when the average temperature is T^o Fahrenheit.

* * * *

1.3.2   Range

The *range* is the set of all answers that you get when you substitute all of the numbers in domain into the function. For instance, the function J in Example 1.4 above has range the set of real numbers between 0 and 100 since the unemployment rate is a percentage.

If $f(x) = x^2$, then $f(2) = 4, f(-1) = 1$ and $f(6) = 36$, so the numbers 4, 1, and 36 are in the range of f. Of course, these are not the only numbers in the range; you could plug in numbers all day and never get the entire range of f. Notice that there are no negative numbers in the range since $x^2 \geq 0$ for any value of x. When a function is given by a formula it is sometimes difficult to find the range. But if we can graph the function then often we can look at its graph to determine the range. The graph of $f(x) = x^2$ is shown below (Figure 1.5). You can see that the range of f consists all real numbers greater than or equal to zero. The graph of this function, a second degree polynomial, is called a parabola. For a discussion on parabolas see chapter 3.

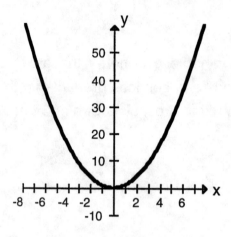

Figure 1.5

Exercises

1. Evaluate each of the following functions at the given value.

   a. $f(t) = \sqrt{5.2 + 3t^2}\,; t = 1.09$

   b. $r(t) = -2.04t^2 + 3.92t - 1; t = 4.7$

   c. $s(x) = \dfrac{1054}{312 - 23x}; x = 67.2$

   d. $A(r) = \dfrac{2.49r}{\sqrt{7.1r - 4.2}}; r = 2$

2. a. Determine the domain of each function in #1.

   b. Graph each function in #1 and determine its range.

In exercises 3 - 8, refer to the graph of the function f given below.

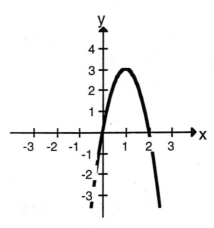

3.   Find $f(0)$ and $f(1)$.

4.   Is $f(1)$ positive or negative?

5.   For what values of x is $f(x)$ positive? negative?

6.   For what values of x is $f(x)$ equal to zero?

7.   What is the domain of f?

8.   What is the range of f?

9.   The following graph shows July rainfall in inches in the city of Atlanta, Ga., from 1990 through 1995.

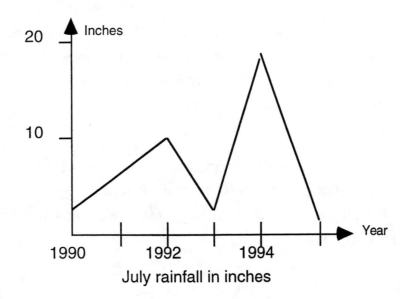

July rainfall in inches

a.   Does this graph represent a function? Explain.

b.   What do the independent and dependent variables represent?

c.   What is the domain? The range?

d.   Write a verbal interpretation of this graph.

1.4   Composite Functions

One way to combine functions to get a new function is to form the *composite function*. Loosely speaking, the composite is a function of a function. To help understand the idea of a composite function let's take a look at a function that relates cricket chirps to temperature in degrees Fahrenheit. This function is

$$K(T) = T - 40,$$

where T is in degrees Fahrenheit and $K(T)$ is the number of chirps in 15 seconds. However, suppose we have temperature gauges that read in Celsius only! Recall from section 1.2 that the function

$$T(C) = \frac{9}{5}C + 32$$

converts Celsius temperatures to Fahrenheit temperatures. We could of course first make the conversion from Celsius to Fahrenheit temperatures than use this result in our chirp function, but it would be more convenient to combine the two functions into one function. This is done by inserting the temperature conversion function into the chirp function. Symbolically we write

$$K[T(C)] = \left[\frac{9}{5}C + 32\right] - 40.$$

Notice that on the right inside the square brackets we have inserted our temperature conversion function. Further, in the square brackets on the left we have inserted the function name $T(C)$. Simplifying the right side of $K[T(C)]$ and renaming, this new function can be written as

$$K(C) = \frac{9}{5}C - 8.$$

In general, if $f(x)$ and $g(x)$ are functions, the *composite function* $f \circ g$ is defined by $f \circ g(x) = f[g(x)]$; that is, the composite is formed if $g(x)$ is substituted for x in $f(x)$ Similarly, $g \circ f$ is defined by $g \circ f(x) = g[f(x)]$; this composite is formed if $f(x)$ is substituted for x in $g(x)$.

Example 1.5

Find the number of chirps you would expect to hear each minute with a temperature of 28 degrees Celsius.

Solution

Let C = 28, then $K(C) = \frac{9}{5}(28) - 8$.

From this, we get

$$K(C) = 50.4 - 8,$$

or

$K(C) = 42.4$ chirps every 15 seconds, or approximately 170 chirps per minute.

As a check on our new function let's solve the same problem as a two step process as suggested above. Solving for Fahrenheit temperature we use the function $T(C) = \dfrac{9}{5}C + 32$.

Then

$$T(C) = \frac{9}{5}(28) + 32,$$

$$T(C) = 50.4 + 32,$$

or

$$T(C) = 82.4 \text{ degrees Fahrenheit.}$$

Now using this result in the chirp function $K(T) = T - 40$ we have

$$K(T) = 82.4 - 40,$$

or

$K(T) = 42.4$ chirps every 15 seconds, or approximately 170 chirps per minute. Thus, the results agree.

* * * *

Group Work

1.   The speed of sound in air is given by the function

$$V(C) = \frac{108}{1 - 0.00366C} \text{ feet per second (fps),}$$

where C is temperature in degrees Celsius. Form the composite function which gives the speed of sound as a function of degrees Fahrenheit. Use this function to find the speed of sound if the temperature is 86 degrees Fahrenheit .

2.   As the population of a community grows, so does the demand for electricity. Often, electricity is produced from coal-burning power plants, and when coal is burned, sulfur dioxide is emitted. Excessive amounts of sulfur in the atmosphere means more acid falls with the precipitation. Suppose that

$f(C)$ is the number of grams of sulfur dioxide emitted from burning C tons of coal, and

$g(P)$ is the number of tons of coal burned to support a population P.
What does the composite function $f[g(P)]$ describe?

* * * *

1.5   Inverse Functions

Some functions have a related function called its *inverse function*; the inverse of a function "undoes" what the function does. The inverse of a function must also be a function. In order for this to happen the original function must be *one-to-one*, i.e., if a and b are two different values in the domain then *f(a)* and *f(b)* must be different. When this is true then the function f will have an *inverse* denoted by f^{-1} and defined by the statement

if $f(x) = y$, then $f^{-1}(y) = x$.

(Note that f^{-1} does not mean $\frac{1}{f}$.) The domain of f is the range of f^{-1} and the range of f is the domain of f^{-1}. Also, the inverse of the inverse of a function is the function itself; that is, $\left(f^{-1}\right)^{-1} = f$.

The relationship between a function and its inverse can be illustrated using composite functions. If we form both composites of a function and its inverse, we get

$f^{-1}[f(x)] = x$ and $f[f^{-1}(y)] = y$.

The functions $T(C)$ and $F(S)$ in section 1.4 above are inverses of each other.

Class Work

1.   Use the definition of inverse function to verify the three properties
$\left(f^{-1}\right)^{-1} = f$, $f^{-1}[f(x)] = x$ and $f[f^{-1}(y)] = y$.

2.   Determine which of the functions whose graphs are shown below have inverses; explain your reasoning.

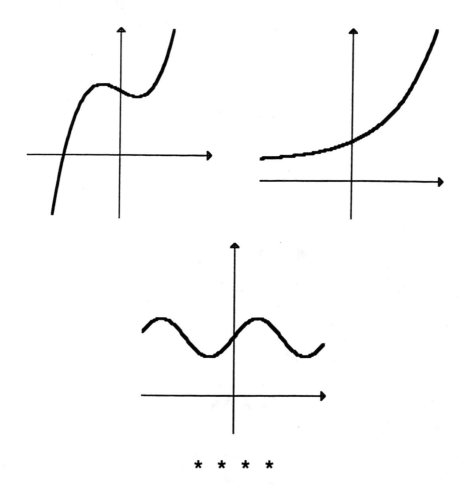

*   *   *   *

Sometimes the definition of the inverse of a function shows us how to find the inverse for some functions; if $f(x)$ is defined by the equation

$$y = f(x),$$

we can determine the equation which defines the inverse f^{-1} if we can solve for x in terms of y to get

$$x = f^{-1}(y).$$

It is not always possible to do this even if the function under consideration has an inverse, but the next two examples illustrate how the technique can be used when it is possible.

Example 1.9

Determine the inverse of f(x) = 4x + 7.

Solution

To find the inverse of this function, replace $f(x)$ by y to get

$$y = 4x + 7.$$

Then solve for x.

$$y - 7 = 4x + 7 - 7 \qquad \text{Add 7 to both sides.}$$
$$y - 7 = 4x \qquad \text{Simplify.}$$
$$\frac{y-7}{4} = \frac{4x}{4} \qquad \text{Divide both sides by 4.}$$
$$x = \frac{y-7}{4} \qquad \text{Simplify and exchange terms across the equal}$$

sign.

So $f^{-1}(y) = x = \dfrac{y-7}{4}$.

As a check, we form the two composites:

$$f^{-1}\left[f(x)\right] = \frac{\left[f(x)\right] - 7}{4} = \frac{(4x + 7) - 7}{4} = x,$$

similarly,

$$f\left[f^{-1}(y)\right] = 4\left[f^{-1}(y)\right] + 7 = 4(\frac{y-7}{4}) + 7 = y.$$

* * * *

Example 1.10

Determine the inverse of $f(x) = 1.37x + 132$.

Solution

Replace $f(x)$ by y,

$$y = 1.37x + 132.5.$$

Solve for x,

$$1.37x = y - 132.5,$$
$$x = \frac{y - 132.5}{1.37}.$$

Hence

$$f^{-1}(y) = \frac{y - 132.5}{1.37}.$$

This time we check only one composite, $f\left(f^{-1}(y)\right)$, and leave the other as

an exercise:

$$f\left(f^{-1}(y)\right) = 1.37\left[\frac{y - 132.5}{1.37}\right] + 132.5 = y.$$

* * * *

Recall from the definition that not all functions have an inverse. If solving for x in terms of y produces two or more x-values, then the "inverse" would not satisfy the definition of a function, so the inverse would not exist. See Example 11.

Example 1.11

Does the function f(x) = x^2 have an inverse?

Solution

Attempting to find the inverse of f(x) = x^2, we replace f(x) with y to get y = x^2. Solving for x gives $x = \pm\sqrt{y}$. If, for example, y = 25, then there are two values for x, 5 and -5, and the "inverse" is not a function. It is, however, a relation.

* * * *

All strictly increasing or strictly decreasing functions have inverses because no two different x-values produce the same y-value. In particular, all non-constant linear functions have inverses.

Class Work

Find the inverse of the function $f(x) = 3x + 7$; verify that $f\left(f^{-1}(y)\right) = y$ and $f^{-1}\left(f(x)\right) = x$.

* * * *

Exercises

For exercises 1-5, determine the inverse of the given linear function, then form both composites to show that $f^{-1}\left(f(x)\right) = x$ and that $f\left(f^{-1}(y)\right) = y$.

1. $f(x) = x$
2. $F(x) = 4 - 3x$
3. $g(x) = \dfrac{x}{5} + 20$
4. $h(x) = 7.314x - 2.001$
5. $f(x) = -5.4x$

The following functions do not have inverses. Explain why.

6. $f(x) = 1 - x^2$

7. $g(x) = x^4$

Determine which of the functions whose graphs are show below have inverses.

8.

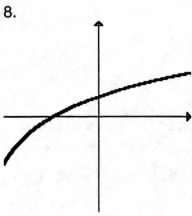

9.

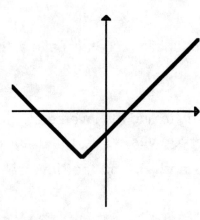

* * * *

CHAPTER TWO
U.S. WATER USAGE AND POPULATION

Introduction

There is a limited amount of fresh water available in the United States, and this water must meet the demand of a growing population. Water is withdrawn from lakes, streams and underground aquifers for basic human needs such as drinking, cooking, and sanitation; it is also used extensively for irrigation as well as industrial and commercial uses. In addition, there are important "in-stream" uses for water: maintenance of water quality, providing habitat for fish and wildlife, navigation, recreational use, and hydroelectric power generation. Also, free flowing rivers and unpolluted lakes are beautiful.

After water is taken out of a river (or lake, stream or reservoir), it may be used and then returned to the river, usually after treatment, possibly in degraded form; or it may be consumed by evaporation, incorporation into products, discharged into the oceans, or otherwise made unavailable for use. The *demand* for water is the amount needed for in-stream use plus the amount withdrawn for all purposes. In the next two chapters we will examine trends in US water use by studying the amount of water withdrawn and its relationship to population. In this chapter we will use linear functions to examine trends for population, total US water withdrawal, and water withdrawal for public use. In the next chapter we see what the trends are in personal consumption and what is the effect from changes in these trends.

Water is withdrawn in the United States for three types of use: for public use, including both urban and rural domestic and commercial purposes, accounts for about 8 - 12% of all US withdrawals; irrigation uses about 33% of the total; industrial uses, including utilities, accounts for 55 - 60%. The approximate shares are shown in the chart below.

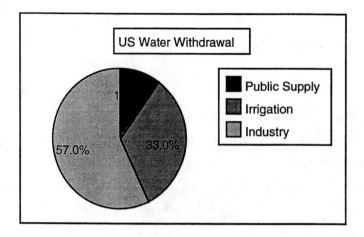

2.1   Population

The following table (Statistical Abstract of the United States) gives resident U.S. population (in millions), for selected years.

Year	Resident U.S. Population (in millions)
1950	152
1960	180
1970	204
1980	227
1990	249
1995	262

Table 2.1

In order to predict water requirements for the future we need to predict the population for future years, i. e., we need to find a formula which can be used to estimate the population for any given year, or at least for any year over some reasonable period of time. The first step in fitting a model to a set of data is to graph the data to get an idea of the kind of model that may be appropriate. We can do this by making points out of the data and plotting these points. We write this data in the format (t, P) called an *ordered pair*. They have as their first coordinate "year" and their second coordinate "population in millions". In order to make the numbers easier to deal with, we adjust the years so that 1950, the first year for which information is provided, corresponds to 0. That is, if t is the variable representing time, then t = 0 means 1950 and more generally, t = number of years since 1950. Also, let P denote the United States population in millions. For example, in 1980 the U.S. population was 227 million; this translates to t = 30, P = 227, and the corresponding point is represented by the ordered pair (30,227). Figure 2.1 shows the data plotted on a (t,P) coordinate system where t is found on the horizontal axis and P is found on the vertical axis.

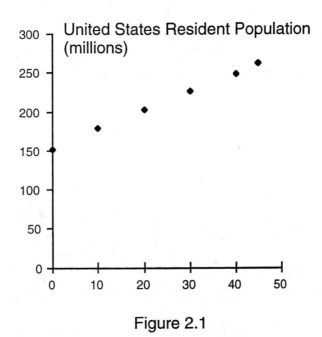

Figure 2.1

The plot of data in figure 2.1 is called a *scatter plot*. Scatter plots are used to analyze the relationships between variables in a set of data. Questions to answer using a scatter plot are:

1.)   Can the relationship between two variables be described by a specific kind of mathematical function?

2.)   If so, what kind?

Figures 2.2 and 2.3 show two types of linear relationships; Figure 2.4 shows a plot that suggest a relationship that is not linear but nicely curved; Figure 2.5 shows a plot that may not reflect a nice relationship between the variables.

In the remainder of this text we will explore different types of functions which may be used to describe certain seta of data.

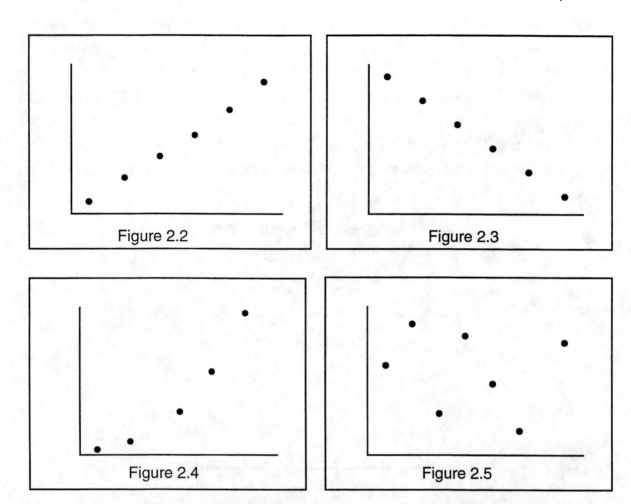

Figure 2.2

Figure 2.3

Figure 2.4

Figure 2.5

2.2   Using a Linear Function

We would like to find a formula whose graph will come close to all the plotted points of Table 2.1 and Figure 2.1. This formula will allow us to estimate the population P for a given year t; that is, we want a formula which describes P as a *function* of t. The points in the Figure 2.1 appear to fall very close to a straight line, so we will use a linear function. We will look at some different linear functions which might be used to approximate the data after a brief review of basic properties of lines and linear functions. More details about linear functions and their properties are provided at the end of this chapter.

A *linear function* is a function with a constant rate of change. It's graph is a straight line, and the *slope* of the line is the *rate of change* of the function. The function

f(x) = mx + b

24

is a linear function whose graph is a line with slope m and vertical intercept (0,b). The slope of the line through two points with coordinates (x_1, y_1) and (x_2, y_2) is given by

$$m = \frac{y_2 - y_1}{x_2 - x_1};$$

the equation of the line can be determined using the *point slope formula*

$$y - y_1 = m(x - x_1).$$

The data points plotted lie close to a straight line but not exactly on one line, so we can't find one linear function which will pass through all of the points. In fact, there are many different linear functions which can be used to approximate the data presented above. We present two possible choices in the following examples.

Example 2.1

Determine the linear function whose graph passes through the first and last data points, that is, through (0,152) and (45, 262). Use this function to estimate the rate at which the population is increasing. Predict the population in the year 2000.

Solution

To determine the function P(t) = mt + b which describes this line, all we have to do is determine the constants m and b. The constant *m* is the slope of the line,

$$m = \frac{262 - 152}{45 - 0} = \frac{110}{45} = 2.44 \text{ (rounded)}.$$

The constant b is the intercept, and since one of the given points is the intercept (0,152), we know $b = 152$. Then

$$P(t) = 2.44t + 152,$$

where P(t) is the population in millions in the 1950 + t.

The slope tells us that the population increases at a rate of 2.44 million per year, i.e., for each one year that passes, the population increases by 2.44 million; the intercept tells us that in 1950 (t = 0) the population was 152 million. The function allows us to estimate the population for any given year. For example, to predict the population in 2000, set t = (2000 - 1950) = 50; P(50) = 274, and the population in the year 2000 is predicted to be 274 million.

* * * *

Example 2.2

Determine the linear function whose graph passes through the last two data points, that is, through (40,249) and (45, 262). Use this function to estimate the rate at which the population is increasing. Predict the population in the year 2000.

Solution

First determine the slope

$$m = \frac{262 - 249}{45 - 40} = \frac{13}{5} = 2.6.$$

Since we don't know the intercept we use the point-slope form for the line. The equation for the line through the point (x_1, y_1) with slope m is

$$y - y_1 = m(x - x_1),$$

or using our variables, the equation of the line through the point (t_1, P_1) with slope m is

$$P - P_1 = m(t - t_1).$$

Either choice of points will produce the same equation, so we will use the first point (40,249) we get

$$P - 249 = 2.6(t - 40)$$

$$P = 2.6\,t + 145$$

$$P(t) = 2.6\,t + 145.$$

The slope, m = 2.6, is the rate of increase; this tells us that the population is increasing at a rate of 2.6 million per year. To predict the population in 2000 evaluate the function when t = 50. Since P(50) = 275, the population in the year 2000 is predicted to be 275 million.

* * * *

An alternative to the *point slope formula* used above is the *slope-intercept form*,

$$y = mx + b,$$

where *m* is the slope and *b* is the vertical intercept, (the y-coordinate of the point where the graph crosses the vertical axis). If the slope is known and *x* and *y* are the coordinates of any point on the line then we can solve for *b*:

$$b = y - mx.$$

Substituting information from Example 2.2, i.e., slope 2.6, x = 40, and y = 249, into the new formula, we have

$$b = 249 - 2.6(40)$$

$$b = 145$$

,

so

$$P(t) = 2.6t + 145.$$

Often it is more convenient to use this latter formula.

Group Work

Determine the linear function passing through the data points for the years:

i)   1980 and 1995;

ii)   1950 and 1990;

iii)   1960 and 1950.

For each of these functions answer the following questions.

1.   What is the predicted rate of change in population?

2.   Predict the population in the year 2000.

3.   When will the population reach 300,000,000?

* * * *

2.3   Choosing the Model

In the previous section we determined several different functions which could be used to approximate the given population data. One way to determine how well a function approximates data is to compare known data, in this case the actual population, with the values which would be predicted according to the function. Table 2.2 indicates the actual values and the predicted values for each of the two functions from the previous section. Predicted values are rounded to the nearest whole number, i.e., nearest million. We will distinguish between the two functions by subscripts,

$$P_1(t) = 2.44t + 152 \text{ and } P_2(t) = 2.60t + 145.$$

Year	t	Population (millions)	$P_1(t)$	$P_2(t)$
1950	0	152	152	145
1960	10	180	176	171
1970	20	204	201	197
1980	30	227	225	223
1990	40	249	250	249
1995	45	262	262	262

Table 2.2

The graphs in Figures 2.6 and 2.7 below show the actual data points and graphs of the functions $P_1(t)$ and $P_2(t)$ respectively.

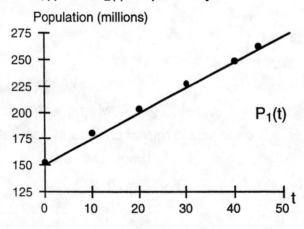

Figure 2.6

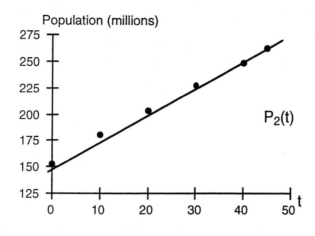

Figure 2.7

Group Work

1.   Answer the questions below for each of the two functions derived in section 2.2. Use the table or the appropriate graph to determine the answers.

        i) For which years does the function over-estimate the actual population? What is the sum of the over-estimates?

        ii) For which years does the function under-estimate the actual population? What is the sum of the under-estimates?

2.   If we decide that the "best" function is the one for which the over-estimate plus the under-estimate is least, which function is best? (Note: use a positive number for both the over-estimate and the under-estimate. Why?)

* * * *

2.3.1   Linear Regression

        We have only considered two possible linear functions to approximate the data. There are many other possibilities, in fact infinitely many possibilities if we don't require that the line passes through any of the data points. Statisticians, mathematicians and scientists using statistics have a definition of "best" function and have derived formulas which can be used to determine which of all possible lines is the best. Their definition of "best" is very close to the one we used above, but instead of taking the absolute value of the differences between predicted and actual values, the square of the differences is used. The technique used for determining this line is called *linear regression,* and the line which is determined is called the *regression line.* Any graphing calculator or software with statistical capabilities can be used to find the regression line. For more information, read the manual, ask your instructor, or experiment. We illustrate the procedure in the next example.

Example 2.3

Determine the regression line for the population data in Table 2.1. Use this function to predict the population in 2000 and to determine when the population will reach 300,000,000.

Solution

Select the statistics mode and then enter the data points. In this case we have six of them. Now calculate the regression line. If the form used is

$$y = ax + b$$

then we get $a = 2.40$ and $b = 154.31$ (rounded to two decimal places). Thus the population function is

$$P(t) = 2.40t + 154.3.$$

Save this function to use for the rest of this chapter.

To predict the population in the year 2000, determine the t-value

$$t = 2000 - 1950 = 50,$$

and evaluate the function at $t = 50$

$$P(50) = 274, \text{ rounded.}$$

We predict a population of 274 million in the year 2000.

To determine when the population will reach 300,000,000 solve the equation $P(t) = 300$. (Recall that the units for population are in millions.)

$$2.40t + 154.3 = 300$$

$$t = 60.7$$

Since $t = 60$ denotes the beginning of the year $1950 + 60 = 2010$, we predict the population will reach 300,000,000 during the year 2010.

When you use your calculator or computer to get the regression equation, you might also see "r = .9992". The number r, called the *correlation coefficient*, is a measure of how close to the regression line the actual data points fall. If all the points are on a line, r = 1 when the line is increasing, and r = -1 when the line is decreasing. In general the value of r lies between -1 and +1. The closer r is to ±1, the better the fit. The value r = .9992 indicates a good fit. Conversely the closer r is to 0 the poorer the fit. In fact, we would say there is no discernible linear relationship when r approaches 0.

* * * *

Group Work

Use the regression equation to answer the following questions.

1.   Predict the population for the current year.
2.   At what rate is the population increasing?
3.   When will the population reach 275 million?
4.   Predict the population in 100 years. Do you think this is reasonable? Explain your answer.
5.   When was the population 0? Is this reasonable? Explain.

* * * *

Your answers in questions 4 and 5 are dealing with what would be a reasonable restriction for the domain of the function. In general, the further away you move from the actual data used, the less accurate your prediction will be. The trend for population in the last 50 years is fairly stable, but it is very different from the trends exhibited earlier; in the future there might be other changes. When you use a function to make a prediction, be aware of the fact that your prediction is valid only if the current trend continues.

Group Work, Continued.
6.   Complete the above work before answering this question. The table below gives population data for the decades 1850 through 1940. Combine this data with the data for the years previously given (1950 - 1995). Plot this new

group of data. Will the plotted points support a linear equation? Explain why or why not.

Year	Resident U.S. Population (in millions)
1850	23
1860	31
1870	39
1880	50
1890	63
1900	76
1910	92
1920	106
1960	123
1940	132

★ ★ ★ ★

2.4 Water Withdrawal

How much water, on the average, does a person in the United States use each year? The answer depends on what we mean by "use". There is direct usage for needs such as drinking water, bathing, etc., which could be estimated by looking at the your water bill. But much water use is indirect, for example, for irrigation, power generation, and industrial usage. In this section we will look at direct personal use as measured by water withdrawal for public supply.

Table 2.3 provides data for selected years describing daily United States water withdrawal for public supply in billions of gallons per day. This includes withdrawal for domestic and commercial use, but it excludes water withdrawn for irrigation, industrial and rural use.

United States Water Withdrawal	
Year	Daily Public Supply (billion gallons per day)
1950	14
1960	21
1970	27
1980	34
1990	41

Table 2.3

Group Work

1.   Plot the points corresponding to the data provided. Use a (t,W) coordinate system with

W = daily withdrawal for public supply (x 10^9 gallons) in year 1950 + t.

2.   Determine the regression line for these data. Round coefficients to two decimal places. This gives a function W(t) which can be used to predict the daily withdrawal for public supply (x 10^9 gallons) in year 1950 + t.

3.   Give a verbal interpretation of the slope.

4.   Predict the annual increase in withdrawal for public supply.

5.   Estimate the current daily and annual withdrawal.

6.   Predict when the daily withdrawal will be 50 billion gallons.

7.   Predict the daily withdrawal for the year 2020.

★ ★ ★ ★

2.5   Linear Functions

In this section we study the different forms for linear functions in more detail.

2.5.1   The slope-intercept equation

A linear function has the form $f(x) = mx + b$, where m and b are constants; its graph is a straight line. A linear function has a constant rate of

33

increase or decrease; this constant rate of change is the *slope* of the line and is represented by m in the expression above. The domain of a linear function consists of all real numbers; the range of a non-constant linear function also consists of all real numbers. (What is the range of a constant linear function?)

The following information was obtained from the Environmental Protection Agency: in 1985 the total landfill capacity in the U.S. was about 250 million tons; furthermore, landfill capacity in the U.S. has been decreasing at a constant rate of about 7.2 million tons per year in recent years. If we let

t = the number of years since 1985

and

C(t) = the landfill capacity in millions of tons in year t,

we can model the landfill capacity in the U.S. with the linear function

C(t) = -7.2t + 250.

The slope m is -7.2; it tells us that when the variable t increases by 1, C will decrease by 7.2 (million tons). You can write the slope as

$$slope = \frac{-7.2}{1}$$

Since the slope is negative, the function is decreasing and the line will fall from left to right. Thus, the slope value tells us how steep the line is and the sign of the slope tells us whether it rises or falls.

The constant b is 250 (million tons); it is the value of the function when t = 0. That is,

C(0) = -7.2(0) + 250 = 250.

The point (0, 250) is called the *vertical intercept* of the line. Graphically, this is the point where the line crosses the vertical axis (see Figure 2.8).

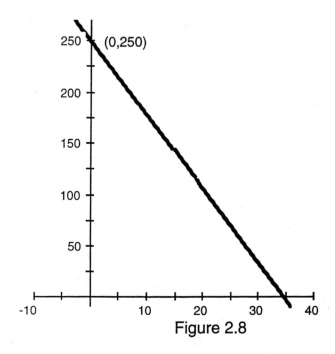

Figure 2.8

In general, the graph of a linear function $f(x) = mx + b$ is a line with slope m and vertical intercept (0,b). If we use an (x,y) coordinate system, the vertical intercept is called the y-intercept. Some times we are interested in the point where the graph crosses the x-axis (x-intercept.) Looking at Figure 2.8 it would appear the we will run out of landfill space in 34 years after 1985 or about the year 2019. Why?

Example 2.4

Find the equation of the line with slope -1.23 and y-intercept (0,4.37).

Solution

Since $m = -1.23$ and $b = 4.37$, the equation is $y = -1.23x + 4.37$.

★ ★ ★ ★

Example 2.5

Find the slope of the line whose equation is $5x + 2y = 4$

Solution

In this case, we know the equation and need to determine the slope. To do this, we will solve for y:

$$2y = -5x + 4$$
$$y = \frac{-5x + 4}{2}$$
$$y = -\frac{5}{2}x + 2.$$

Now the equation is in slope-intercept form, and we see that the slope is $-\frac{5}{2}$.

* * * *

2.5.2   The slope formula

The slope can be thought of as the ratio of the vertical change to the horizontal change between two points. (Some times this is said to be the *rise over run*.) If the two points are (x_1, y_1) and (x_2, y_2), the vertical change is $y_2 - y_1$ (the rise) and the horizontal change is $x_2 - x_1$ (the run). Hence the formula

$$m = \frac{y_2 - y_1}{x_2 - x_1}$$

gives you a way to determine the slope of the line through these two points. See Figure 2.9.

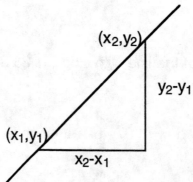

Figure 2.9

Example 2.6

Determine the slope of the line which passes through the points (5, 1.00) and

(15, 1.50).

Solution

Using the slope formula

$$m = \frac{y_2 - y_1}{x_2 - x_1} = \frac{1.5 - 1.0}{15 - 5} = \frac{0.5}{10} = 0.05,$$

the slope m is .05.

* * * *

Another way of describing the slope is by the expression $m = \frac{\Delta y}{\Delta x}$ where Δy is the vertical change and Δx is the horizontal change. The symbol Δ is read as "d*elta*", and in mathematics often means change. Thus Δy is read as delta-y and Δx as delta-x and $\Delta y = y_2 - y_1$ and $\Delta x = x_2 - x_1$.

Given a set of data, we can determine whether the corresponding points fall nearly on a straight line by determining the slopes of the line segments between successive points. If these slopes are constant or almost constant then the data can be approximated by a linear function.

Example 2.7

Determine whether the following data which gives the minimum wage in dollars from 1955 to 1970 can be approximated by a linear function.

YEAR	1955	1960	1965	1970
WAGE	$.75	$1.00	$1.25	$1.50

Solution

If the corresponding points are (year, wage) then we compute slopes between successive points:

37

$$\frac{1.00 - .75}{1960 - 1955} = .05;$$

$$\frac{1.25 - 1.00}{1965 - 1960} = .05;$$

$$\frac{1.50 - 1.25}{1970 - 1965} = .05.$$

The slopes are constant so the corresponding points lie on a single line. (This is unusual in real life; data may be approximated by a linear function but rarely do all the data points lie on the same line.)

* * * *

Class Work

1. Determine the slope and the y-intercept of the line
    $y = .45t + 1.2.$

2. Determine whether or not the following points lie on a single line.

x	5	9	13	17
y	120	167	214	261

3. Find the slope of the line passing through the points (-1.2, 3.5) and (2.3, 4.7).

* * * *

The following are some useful facts about the slope of a line (See Figure 2.10):

1. if the slope is positive, the linear function is increasing; the graph rises from left to right;

2. if the slope is negative, the linear function is decreasing; the graph falls from left to right;

3. if the slope is zero, then the linear function is constant; the graph is a horizontal line;

4. if the slope is undefined, the line is vertical; its equation is $x = a$ where a is the x-intercept.

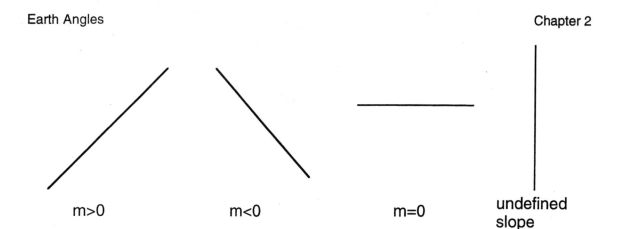

m>0   m<0   m=0   undefined slope

Figure 2.10

Two additional properties of slopes involve parallel and perpendicular lines. Parallel lines have equal slopes; the slopes of perpendicular lines are negative reciprocals to each other. In other words, if the lines have slope m_1 and m_2 respectively, then they are parallel if $m_1 = m_2$ and perpendicular if $m_2 = -\dfrac{1}{m_1}$. (Note that this equation doesn't make sense if either m_1 or m_2 is equal to zero. However, we then know that one line is vertical, the other line is horizontal, and so the lines are perpendicular.)

2.5.3   The point-slope equation.

Suppose that we want to find the equation of a line when one point on the line and its slope are known. If we let (x_1, y_1) represent the known point and (x, y) represent any other point on the line then the slope is

$$\frac{y - y_1}{x - x_1} = m$$
.

Now, multiply both sides of the equation by x - x_1 to get

$$y - y_1 = m(x - x_1)$$.

This is called the *point-slope equation* of a line.

Example 2.8

Write the equation of the line which passes through the point (-2, 5) and has slope 2.

Solution

Set $x_1 = -2$, $y_1 = 5$, and $m = 2$ to get
$$y - 5 = 2(x - (-2)).$$

Simplify and solve for y, and the equation is
$$y = 2x + 9.$$

* * * *

Now suppose you know two points and need to find the equation for the line through them. You can first compute the slope using the slope formula, then use either of the two known points in the point-slope formula.

Example 2.9

Find the equation of the line passing through the points (-3,7) and (2, 5).

Solution

The slope is
$$\frac{5 - 7}{2 - (-3)} = \frac{-2}{5} = -\frac{2}{5}.$$

Then, using the point (-3, 7), the equation is
$$y - 7 = -\frac{2}{5}(x - (-3)).$$

Simplify and solve for y to get
$$y = -\frac{2}{5}x + \frac{29}{5}.$$

Notice that the same equation results if the point (2,5) is used; the equation
$$y - 5 = -\frac{2}{5}(x - 2)$$

also simplifies to
$$y = -\frac{2}{5}x + \frac{29}{5}.$$

* * * *

Class Work

4.   Find the equation of the line with slope -2 passing through the point (1, -5).

5.   Find the equation of the line passing through the points (4.12, 11.23) and (3.08, -7.45).

★ ★ ★ ★

Exercises

1.   Determine the slope and y-intercept of the line whose equation is $2x + 3y - 5 = 0$.

2.   Find the equation of the line passing through the point (1.7, 2.1) with slope -2.

3.   Find the equation of the line passing through the points (20, 18.5) and (48, 19.4).

4.   Find the equation of each line passing through (12, 2.4) which is:

      a.   parallel to the line $y = 3x - 7$;

      b.   perpendicular to the line $y = -5x - 6$.

5.   Estimate the slope of the line whose graph is shown below and find an equation for that line.

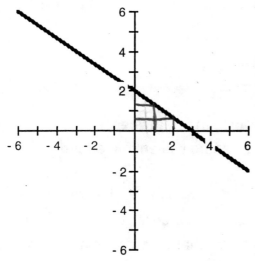

6.   Graph the function $f(x) = .002x - 12$ using your calculator. Be sure to show both the x-intercept and the y-intercept.

7.   Is the function $f(x) = -10.2x + 19$ increasing or decreasing? Explain.

8.   Determine whether each of the following sets of data can be approximated by a linear function. If so, write the function .

a.

s	3	4	5	6	7
t	1	1.2	1.4	2.6	1.8

b.

x	0	1	2	3	4
w	3	5	6	8	11

c.

a	5	8	10	15	18
m	150	120	100	50	20

9.   Scientists have determined that in 1939, the level of carbon dioxide in the air was 280 parts per million and that each year the level increases by about 1.44 ppm. Write the linear function that can be used to model carbon dioxide concentration in the air as a function of time in years since 1939. Then use your function to determine the CO_2 level this year.

10.   Explain why each of the useful facts about the slope of a line is true. (See Figure 2.6.)

* * * *

2.6   Systems Of Linear Equations And Matrices

A *linear equation in n variables* can be written in the form

$$a_1x_1 + a_2x_2 + ...a_nx_n = c,$$

where $x_1, x_2,..., x_n$ are the variables and $a_1, a_2,..., a_n$ and c are constants. A *system of m linear equations in n variables* is a collection of m linear equations each containing the same n variables; for short, we call this an m x n system. Some examples are given below.

Example 2.10

      (a) $3x - y = 4$

         $-2x + y = 3$

This system has two linear equations containing two variables x and y; it is 2×2 system.

      (b) $x - y + z = 8$

          $x + 2y - z = -5$

         $2x + 3y + z = 6$

This has three linear equations containing three variables x, y and z; it is a 3×3 system.

      (c) $x + y + z = 5$

         $3x - y - z = -1$

This one has two linear equations containing three variables x, y, and z; it is a 2×3 system.

      (d) $x + y = 2$

         $y + 3 = z$

        $w - 4 = 0$

The variables here are x, y, z, and w; this is a 3 x 4 system.

* * * *

A *solution* of a system of equations consists of values for the variables that make all of the equations in the system true. For example, x = 7, y = 17, is a solution of the system in Example 1a because

        3(7) - 17 = 4 and -2(7) + 17 = 3.

2.6.1   Solving Linear Systems By The Elimination Method

We will focus on methods of finding solutions to 2 x 2 and 3 x 3 systems of linear equations. One way to solve a system of linear equations is known as the method of *elimination of variables*. The idea behind this method is to eliminate variables by replacing the original equations in the system with equivalent equations until one equation has only one variable left. The following operations are used to produce equivalent equations:

1. Interchange two equations in the system.
2. Replace any equation in the system by a non zero multiple of that equation.
3. Replace any equation in the system by the sum (or difference) of that equation and a multiple of any other equation in the system.

Example 2.11

Use the method of elimination to solve the 2 x 2 system
$$x + y = 5$$
$$3x - 4y = 1.$$

Solution

First we will eliminate the variable y in the second equation. Multiply the first equation by 4 so that the coefficients of y in the two equations are opposites of each other and add this equation to the second equation:

$$4(x+y=5) \qquad 4x+4y=20$$
$$\text{or}$$
$$+(3x-4y=1) \qquad +(3x-4y=1),$$

so the new second equation is
$$7x = 21,$$
and hence
$$x = 3.$$

To find y, substitute $x = 3$ into either of the original equations; if we choose the first equation, we get
$$3 + y = 5,$$
so
$$y = 2.$$

Thus the solution to the system is $x = 3, y = 2$.

Geometrically, this solution gives the coordinates of the point of intersection of the graph of the two equations. See Figure 2.11.

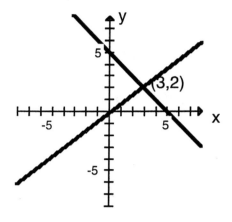

Figure 2.11

* * * *

Example 2.12

Solve the system
$$3x - y = 4$$
$$-6x + 2y = 10.$$

Solution

We eliminate x in the second equation by multiplying the first equation by 2 and adding this to the second equation:
$$6x - 2y = 8$$
$$+(-6x + 2y = 10)$$

which yields
$$0 = 18.$$

Since this is not possible, that means that the system has no solution. Geometrically, the graphs of these two equations are parallel lines. See Figure 2.12.

45

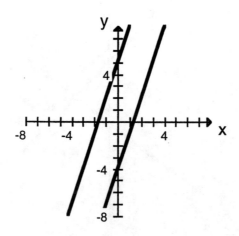

Figure 2.12

* * * *

Class Work

Solve the following systems of linear equations by the elimination method and illustrate your answers with a graph.

1. $x + 3y = 1$
   $2x - y = 2$

2. $2x - 3y = 1$
   $-4x + 6y = 3$

3. $-3x - y = -2$
   $6x + 2y = 4$

* * * *

Example 2.13

Solve the 3 x 3 system
$$x - y + z = 8$$
$$x + 2y - z = -5$$
$$2x + 3y + z = 6.$$

Solution

There's more to do here because there are more variables. First we eliminate z from the second equation by adding equations 1 and 2 to get
$$2x + y = 3.$$
Next we eliminate z from the third equation by adding -1 times equation 1 to equation 3; this gives
$$x + 4y = -2.$$

46

Now we can solve the resulting 2 x 2 system

$$2x + y = 3$$
$$x + 4y = -2$$

as in the previous example. Eliminate x by adding -2 times the second equation to the first equation:

$$2x + y = 3$$
$$+(-2x - 8y = 4).$$

This yields

$$-7y = 7$$

so

$$y = -1.$$

Substitute $y = -1$ into $x + 4y = -2$ (or into $2x + y = 3$),

$$x + 4(-1) = -2$$

or

$$x = 2.$$

Finally substitute $x = 2$ and $y = -1$ into any of the three original equations to find z. If we choose the first, we get

$$2 - (-1) + z = 8,$$

so

$$z = 5.$$

Therefore, the solution is

$$x = 2, y = -1, z = 5.$$

You can check these answers by substituting into all of the original equations in the system.

* * * *

Class Work

      Solve the following 3 x 3 system using the elimination method.

$$x + 2y + z = 4$$
$$x + 3y - 5z = -2$$
$$2x - y - 2z = 4.$$

* * * *

Exercises

Solve each of the following systems of linear equations by the elimination method. If the system has no solution explain why not.

1. $x + 3y = 1$
    $2x - 3y = 2$

2. $4x + y = 5$
    $8x + 3y = 2$

3. $x - y = 1$
    $2x + 3y = -2$

4. $x - y = 3$
    $2x + 3y = 1$

5. $2x + 3y = 1$
    $4x + 6y = 3$

6. $2x + 3y = 1$
    $4x + 6y = 2$

7. $x + 2y + z = 3$
    $x + 3y + 4z = 6$
    $2x + 4y + 3z = 7$

8. $2x + y + z = 7$
    $2x + y - z = 5$
    $2x + 4y + 3z = 12$

9. $2x + y + 3z = 1$
    $3x - y - 2z = 3$
    $4x - y + 3z = 5$

10. $2x + 3y - z = 0$
     $3x - 2y + z = 7$
     $4x + 6y - 3z = 5$

Use the Zoom and Trace features or the Intersection feature of your graphing calculator or computer software to approximate the solution of each of the following systems of linear equations.

11. $0.1x + y = 13$
     $-1.4x + y = 5$

12. $0.2x + 0.3y = -2.2$
     $2.2x - 2.4y = -6$

13. $x - 0.2y = 2.6$
     $x + 1.5y = 6$

* * * *

2.6.2   Matrices

When we solve a system of linear equations using the elimination method, we perform computations with the constants; the names of the variables play no important role. Therefore we can simplify the process by omitting the variables. Consider the system of equations

$$x - y + z = -4$$
$$2x - 3y + 4z = 15$$
$$5x + y - 2z = 12.$$

If we omit the variables x, y and z, we can write the coefficients in order

48

$$\begin{array}{rrr} 1 & -1 & 1 \\ 2 & -3 & 4 \\ 5 & 1 & -2 \end{array}$$

This is an example of a matrix. We usually enclose a matrix in parentheses or brackets. If we name the matrix in our example with the letter, A, then it is written

$$A = \begin{bmatrix} 1 & -1 & 1 \\ 2 & -3 & 4 \\ 5 & 1 & -2 \end{bmatrix}.$$

In this section we provide some terminology, definitions and properties of matrices; then we will show how matrices can sometimes be used to solve systems of linear equations.

A *matrix* is a rectangular array of numbers. Its *size* or *dimension* is designated by the number of rows and columns it has; for example, a matrix with three rows and four columns is a 3 x 4 matrix, and in general a matrix with m rows and n columns is said to have dimension m x n. The numbers in the matrix are called the *entries*; the number in the ith row and j^{th} column is the i,j entry. A matrix having a single row is called a *row matrix*; one with a single column is called a *column matrix*, and a matrix with the same number of rows and columns is called a *square matrix*. Matrices are *equal* if they have the same dimension and the corresponding entries are equal.

An example of a 3 x 4 matrix is

$$\begin{bmatrix} 2 & 0 & 1 & 3 \\ -1 & 2 & 4 & 5 \\ 5 & \frac{3}{4} & \sqrt{2} & 0 \end{bmatrix};$$

its 2,3 entry is 4, its 1,2 entry is 0, etc. The general form for an $m \times n$ matrix is

$$\begin{bmatrix} a_{11} & a_{12} & \cdots & a_{1n} \\ a_{21} & a_{22} & \cdots & a_{2n} \\ . & . & \cdots & . \\ a_{m1} & a_{m2} & \cdots & a_{mn} \end{bmatrix};$$

the i,j entry is a_{ij} (the term in row i, column j).

We can do arithmetic with matrices provided that they are of the right size. Two matrices of the same dimension can be added together or subtracted from each other by adding or subtracting the corresponding entries, as shown in the next examples.

Example 2.14

$$\begin{bmatrix} 2 & 1 & 0 \\ -1 & 4 & 5 \end{bmatrix} + \begin{bmatrix} 5 & -2 & 3 \\ 4 & 0 & 7 \end{bmatrix} = \begin{bmatrix} 7 & -1 & 3 \\ 3 & 4 & 12 \end{bmatrix}$$

* * * *

Example 2.15

$$\begin{bmatrix} 5 & 1 \\ 0 & 3 \\ -1 & 4 \\ 0 & 2 \end{bmatrix} - \begin{bmatrix} 3 & 2 \\ -4 & 0 \\ -1 & 3 \\ 5 & -2 \end{bmatrix} = \begin{bmatrix} 2 & -1 \\ 4 & 3 \\ 0 & 1 \\ -5 & 4 \end{bmatrix}.$$

* * * *

Any matrix can be multiplied by any real number; multiply each entry by that number. This operation, called *scalar multiplication*, is illustrated in the next example.

Example 2.16

$$5 \begin{bmatrix} 1 & 0 & -2 \\ 0 & 3 & 1 \end{bmatrix} = \begin{bmatrix} 5 & 0 & -10 \\ 0 & 15 & 5 \end{bmatrix}.$$

* * * *

The operation on matrices we will use for solving systems of equations is *matrix multiplication*. If two matrices are the right size, then they can be multiplied to produce a new matrix. In the first example, we multiply a 1 x 4 row matrix A with a 4 x 1 column matrix B to get a 1 x 1 product AB.

Example 2.17

$$\begin{bmatrix} 1 & 3 & 2 & 4 \end{bmatrix} \begin{bmatrix} 2 \\ 5 \\ 3 \\ 1 \end{bmatrix} = [1 \cdot 2 + 3 \cdot 5 + 2 \cdot 3 + 4 \cdot 1] = [27].$$

50

Here $A = \begin{bmatrix} 1 & 3 & 2 & 4 \end{bmatrix}$ is 1 x 4, $B = \begin{bmatrix} 2 \\ 5 \\ 3 \\ 1 \end{bmatrix}$ is 4 x 1, and the product $AB = [27]$ is 1x1.

* * * *

Example 2.17 is a model for how to multiply a $1 \times k$ row matrix by a $k \times 1$ column matrix to produce a 1×1 matrix: just multiply the corresponding entries and sum. In general, if A is $m \times k$ and B is $k \times n$ then AB is $m \times n$. The product is defined by multiplying the ith row of A with the jth column of B to produce the i,j entry of AB. If the number of columns of A does not equal the number of rows of B then the product AB is not defined. The next example illustrates the techniques of matrix multiplication.

* * * *

Example 2.18

If $A = \begin{bmatrix} 1 & 3 \\ 2 & -1 \\ 0 & 4 \end{bmatrix}$ and $B = \begin{bmatrix} 5 & 1 & 3 & -2 \\ -1 & 2 & 0 & 1 \end{bmatrix}$ find AB and BA

Solution

First we compute
$$AB = \begin{bmatrix} 1 & 3 \\ 2 & -1 \\ 0 & 4 \end{bmatrix} \begin{bmatrix} 5 & 1 & 3 & -2 \\ -1 & 2 & 0 & 1 \end{bmatrix}.$$

To determine the i,j entry, multiply row i of the first matrix with row j of the second matrix to get

$$AB = \begin{bmatrix} 1\cdot5+3(-1) & 1\cdot1+3\cdot2 & 1\cdot3+3\cdot0 & 1(-2)+3\cdot1 \\ 2\cdot5+(-1)(-1) & 2\cdot1+(-1)\cdot2 & 2\cdot3+(-1)0 & 2(-2)+(-1)\cdot1 \\ 0\cdot5+4(-1) & 0\cdot1+4\cdot2 & 0\cdot3+4\cdot0 & 0(-2)+4\cdot1 \end{bmatrix} = \begin{bmatrix} 2 & 7 & 3 & 1 \\ 11 & 0 & 6 & -5 \\ -4 & 8 & 0 & 4 \end{bmatrix}.$$

The product BA is not defined since B is 2 x 4 and A is 3 x 2.
* * * *

Example 2.19

If $A = \begin{bmatrix} 1 & 2 \\ 3 & 4 \end{bmatrix}$ and $X = \begin{bmatrix} x \\ y \end{bmatrix}$ determine AX.

Solution

$$AX = \begin{bmatrix} 1 & 2 \\ 3 & 4 \end{bmatrix}\begin{bmatrix} x \\ y \end{bmatrix} = \begin{bmatrix} 1 \cdot x + 2 \cdot y \\ 3 \cdot x + 4 \cdot y \end{bmatrix} = \begin{bmatrix} x + 2y \\ 3x + 4y \end{bmatrix}$$

* * * *

Class Work

Multiply

$$\begin{bmatrix} 1 & 2 & 3 \\ 3 & -1 & 0 \\ 0 & 1 & -2 \end{bmatrix}\begin{bmatrix} 4 & 0 & 1 \\ 1 & 3 & 2 \\ 5 & -1 & 6 \end{bmatrix}$$

* * * *

Exercises

Perform the indicated operations if possible.

1. $\begin{bmatrix} 2 & 0 \\ 4 & 6 \end{bmatrix} + \begin{bmatrix} 4 & 0 \\ 5 & 1 \end{bmatrix}$

2. $\begin{bmatrix} 2 & 1 \\ 4 & 3 \end{bmatrix} - \begin{bmatrix} 4 & 5 \\ 0 & 1 \end{bmatrix}$

3. $\begin{bmatrix} 1 & 3 & 2 \\ 4 & 1 & 0 \end{bmatrix} - \begin{bmatrix} -1 & 3 & -2 \\ 3 & 4 & 2 \end{bmatrix}$

4. $\begin{bmatrix} 1 & 3 & 0 \\ 4 & 1 & 0 \end{bmatrix} + \begin{bmatrix} 1 & 3 \\ 3 & 4 \end{bmatrix}$

5. $16\begin{bmatrix} 1 & -1 & .25 & -.125 \end{bmatrix}$

6. $6\begin{bmatrix} -2 & 1 \\ 5 & 0 \end{bmatrix}$

7. $\begin{bmatrix} 2 & 0 \\ 4 & 6 \end{bmatrix}\begin{bmatrix} 4 & 0 \\ 5 & 1 \end{bmatrix}$

8. $\begin{bmatrix} 2 \\ 5 \end{bmatrix}\begin{bmatrix} 2 & 5 \end{bmatrix}$

9. $\begin{bmatrix} 1 & 3 & 2 \\ 4 & 1 & 0 \end{bmatrix} \begin{bmatrix} -2 & 3 \\ 1 & 0 \\ 1 & 2 \end{bmatrix}$ 10. $\begin{bmatrix} -2 & 3 \\ 1 & 0 \\ 1 & 2 \end{bmatrix} \begin{bmatrix} 1 & 3 & 2 \\ 4 & 1 & 0 \end{bmatrix}$

* * * *

2.6.3   Using Matrices to Solve Systems of Equations

We return to the system of equations (Section 2.6.2)

$$x - y + z = -4$$
$$2x - 3y + 4z = 15$$
$$5x + y - 2z = 12.$$

The coefficients form a 3 x 3 matrix

$$A = \begin{bmatrix} 1 & -1 & 1 \\ 2 & -3 & 4 \\ 5 & 1 & -2 \end{bmatrix}$$

called the coefficient matrix of our system of equations; there is one row for each equation in the system and one column for each of the three variables. The variables in the system of equations are represented as a column matrix of dimension 3×1

$$X = \begin{bmatrix} x \\ y \\ z \end{bmatrix},$$

and the constants are represented by the 3×1 column matrix

$$B = \begin{bmatrix} -4 \\ -15 \\ 12 \end{bmatrix}.$$

The system of equations can now be written as a matrix equation. We will write the coefficient matrix times the matrix of variables on the left side and the matrix of constants on the right side,

$$\begin{bmatrix} 1 & -1 & 1 \\ 2 & -3 & 4 \\ 5 & 1 & -2 \end{bmatrix} \begin{bmatrix} x \\ y \\ z \end{bmatrix} = \begin{bmatrix} -4 \\ -15 \\ 12 \end{bmatrix}$$

or

$$AX = B$$

This gives all the information we had in the original system of equations. If we multiply the matrices on the left, we get the 3 x 1 matrix

$$AX = \begin{bmatrix} x+y+z \\ 2x-3y+4z \\ 5x+y-2z \end{bmatrix}.$$

Then the matrix equation $AX = B$ is

$$\begin{bmatrix} x-y+z \\ 2x-3y+4z \\ 5x+y-2z \end{bmatrix} = \begin{bmatrix} -4 \\ 15 \\ 12 \end{bmatrix}.$$

In order for this equation to be true, the matrix on the left must be the same as the matrix on the right. That is,

$$x-y+z = -4, \ 2x-3y+4z = 15, \text{ and } 5x+y-2z = 12$$

or

$$x-y+z = -4$$
$$2x-3y+4z = 15$$
$$5x+y-2z = 12.$$

Thus the matrix equation $AX = B$ is equivalent to the original system of equations.

Class Work

Write each system of equations as a matrix equation and use matrix multiplication to show that your matrix equation is equivalent to the system.

1. $x-y+3z = 7$ 2. $-4x+y-2z = 7$
         $2x+y+z = 4$ $-3x-y+2z = 3$
     $3x-2y+2z = 10$ $5x+y-3z = 4$

★ ★ ★ ★

The 3 x 3 identity matrix is a special square matrix of the form

$$\begin{bmatrix} 1 & 0 & 0 \\ 0 & 1 & 0 \\ 0 & 0 & 1 \end{bmatrix}.$$

The *n x n identity matrix* is a square matrix with 1's extending along the main diagonal and 0's elsewhere. An identity matrix behaves kind of like the number "1"; if I is an identity matrix and A is a square matrix of the same dimension then AI = IA = A.

Example 2.20

$$\begin{bmatrix} 1 & 5 & 3 \\ 2 & 1 & 4 \\ 3 & 5 & 6 \end{bmatrix}\begin{bmatrix} 1 & 0 & 0 \\ 0 & 1 & 0 \\ 0 & 0 & 1 \end{bmatrix} = \begin{bmatrix} 1 & 5 & 3 \\ 2 & 1 & 4 \\ 3 & 5 & 6 \end{bmatrix}$$

and

$$\begin{bmatrix} 1 & 0 & 0 \\ 0 & 1 & 0 \\ 0 & 0 & 1 \end{bmatrix}\begin{bmatrix} 1 & 5 & 3 \\ 2 & 1 & 4 \\ 3 & 5 & 6 \end{bmatrix} = \begin{bmatrix} 1 & 5 & 3 \\ 2 & 1 & 4 \\ 3 & 5 & 6 \end{bmatrix}$$

∗ ∗ ∗ ∗

The multiplicative inverse of a square matrix, A, is a matrix, A^{-1}, for which $AA^{-1} = I$.

For example, the inverse of the matrix

$$A = \begin{bmatrix} 4 & 1 & -2 \\ -3 & -1 & 2 \\ 5 & 1 & -3 \end{bmatrix}$$

is

$$A^{-1} = \begin{bmatrix} 1 & 1 & 0 \\ 1 & -2 & -2 \\ 2 & 1 & -1 \end{bmatrix}.$$

Check as an exercise that $AA^{-1} = I$ and $A^{-1}A = I$.

Only square matrices have inverses, and not all of these do. Your graphing calculator or computer algebra package has the capability of finding the inverse of a matrix (if it has one).

Class Work

Use your calculator or computer to find the inverse of each matrix then use matrix multiplication to show that $AA^{-1} = I$ and $A^{-1}A = I$.

1. $A = \begin{bmatrix} 1 & 1 \\ -2 & 3 \end{bmatrix}$

55

2. $A = \begin{bmatrix} 1 & 1 & 0 \\ -1 & 3 & 4 \\ 0 & 4 & 3 \end{bmatrix}$

* * * *

Now we have enough information to use matrices to solve a system of equations. We return to the system

$$x - y + z = -4$$

$$2x - 3y + 4z = 15$$

$$5x + y - 2z = 12$$

and its equivalent matrix equation

$$\begin{bmatrix} 1 & -1 & 1 \\ 2 & -3 & 4 \\ 5 & 1 & -2 \end{bmatrix} \begin{bmatrix} x \\ y \\ z \end{bmatrix} = \begin{bmatrix} -4 \\ -5 \\ 12 \end{bmatrix}.$$

Recall that

$$A = \begin{bmatrix} 1 & -1 & 1 \\ 2 & -3 & 4 \\ 5 & 1 & -2 \end{bmatrix}, \quad X = \begin{bmatrix} x \\ y \\ z \end{bmatrix}, \text{ and } B = \begin{bmatrix} -4 \\ -5 \\ 12 \end{bmatrix}.$$

and the matrix equation is AX = B. To solve this equation we need to multiply each side of the equation on the left by A^{-1}

$$A^{-1} \cdot AX = A^{-1} \cdot B$$

$$IX = A^{-1} \cdot B$$

$$X = A^{-1} \cdot B.$$

Thus to solve the matrix equation AX = B, we simply need to find the matrix inverse A^{-1} and then find the product $A^{-1}B$. This is easy to do using a calculator or computer software. First, set the dimensions of A and B, type in the entries, then compute the product $A^{-1}B$ to see the solution matrix.. Note that you do not need to enter the matrix X; the product $A^{-1}B$ will equal X. In the example above, we have

$$A^{-1} = \begin{bmatrix} -.4 & .2 & .2 \\ -4.8 & 1.4 & .4 \\ -3.4 & 1.2 & .2 \end{bmatrix} \text{ and } A^{-1}B = \begin{bmatrix} 1 \\ 3 \\ -2 \end{bmatrix};$$

hence the solution to the system of equations is x = 1, y = 3, and z = -2.

Example 2.21

Use the matrix inverse method to solve the system

$$4x + y = 7$$
$$2x + 3y = -1.$$

Solution

The coefficient matrix is the 2 x 2 matrix

$$A = \begin{bmatrix} 4 & 1 \\ 2 & 3 \end{bmatrix},$$

the matrix of constants is the 2 x 1 matrix

$$B = \begin{bmatrix} 7 \\ -1 \end{bmatrix};$$

and the solution matrix is

$$X = A^{-1}B = \begin{bmatrix} 2.2 \\ -1.8 \end{bmatrix}.$$

Thus the solution to the original system of equations is

x= 2.2, y = -1.8.

* * * *

Example 2.22

Use the matrix inverse method to solve the system

$$4x + y - 2z = 7$$
$$-3x - y + 2z = 3$$
$$5x + y - 3z = 4.$$

Solution

The coefficient matrix is the 3 x 3 matrix

$$A = \begin{bmatrix} 4 & 1 & -2 \\ -3 & -1 & 2 \\ 5 & 1 & -3 \end{bmatrix},$$

the constant matrix is the 3 x 1 matrix

$$B = \begin{bmatrix} 7 \\ 3 \\ 4 \end{bmatrix}.$$

The solution matrix is

$$X = A^{-1}B = \begin{bmatrix} 10 \\ -7 \\ 13 \end{bmatrix}$$

which means that the solution to the original system is

x = 10, y = -7, z = 13.

* * * *

Exercises

Solve the following systems of equations using matrix inverses; use your calculator or computer.

1. $x + 2y = 2$
   $3x + 2y = 5$

2. $3x + 7y = 4$
   $x + 4y = 5$

3. $7x + y = 1$
   $3x + 4y = -1$

4. $11x + y = 3$
   $3x + y = 7$

5. $3x + 2y = 8$
   $x + 4y = 5$

6. $15x + 12y = 10$
   $21x + 4y = -13$

7. $8x + 3y = 17$
   $-11x + 2y = 3$

8. $x - 3y - 2z = 11$
   $2x + 8y + 9z = 3$
   $x + 5y + 6z = 13$

9. $x - y + 3z = 2$
   $3x - y + 7z = 3$
   $x + y + 6z = -5$

10. $x + 7y + 4z = -3$
    $-3x - y + 16z = 7$
    $2x + 4y - z = 1$

11. $x + 3y + 7z = 4$
    $3x + 4y + 7z = 5$
    $2x + 6y + 9z = -8$

12. $5x + 2y - z = 4$
    $4x - 3y + 2z = 5$
    $3x + 7y - 5z = 8$

13. $100x + 10y + z = 13.9$
    $400x + 20y + 2z = 13.5$
    $900x + 30y + z = 13.4$

* * * *

58

CHAPTER THREE
PER CAPITA WATER USE

Introduction

In this chapter, we use quadratic and rational functions to study trends in per capita water withdrawal, and learn the properties of these functions.

3.1 Per Capita Water Withdrawal

The amount of water we use is increasing each year. In the last chapter we saw that water withdrawal for public supply is increasing over time and we modeled that phenomenon with a linear function with positive slope. It isn't surprising that we use more water each year since the US population is increasing. What isn't yet clear is whether each person is, on the average, using more water, less water or the same amount of water each year. The quantity which will describe this is the *per capita water withdrawal* (for public supply), i.e., the withdrawal per person which is

$$\text{per capita water withdrawal} = \frac{\text{withdrawal}}{\text{population}}.$$

We let C(t) designate the per capita daily water withdrawal for public supply in the year 1950 + t. The units are important; water withdrawal is measured in billion (10^9) gallons per day and population is measured in million (10^6) people. Then the per capita rate is $\dfrac{\text{billion gallons per day}}{\text{million persons}}$, which simplifies to

$$\frac{10^9 \text{ gallons per day}}{10^6 \text{ persons}} = \frac{10^9 \text{ gallons}}{10^6 \text{ persons} * \text{day}} = 10^3 \text{ gallons per person per day}$$

or one thousand gallons per person per day. (Note: *per* stands for divide.)
Thus

$$C(t) = \frac{W(t)}{P(t)} \times 10^3 \text{ gallons per person per day in year } 1950+t.$$

C(t) is a *rational function*, that is, the quotient of two polynomial functions, in this case two linear functions. In the exercises below we will examine the function C(t) and determine how much more, or less, water each person has been using each year.

Group Work

In the work below, use the functions (the regression equations) for P(t) and W(t) developed in section 2.3 and 2.4. Give answers for C(t) rounded to the nearest gallons per person per day.

1.   Determine the per capita daily withdrawal for 1968, 1983 and 1997.

2.   Graph the function $C(t) = \dfrac{W(t)}{P(t)} \times 10^3$. Your graph should show per capita withdrawal for the years 1950 to 2010.

3.   For the period 1950 to 2010, when was per capital withdrawal greatest? When was the least? What were these greatest and least withdrawals?

4.   What trends do you see looking at the data? What trends do you see looking at the graph?

*   *   *   *

The rational function C(t) has a *horizontal asymptote*. In this case the values of C(t) increase to a fixed number as t gets larger. We can estimate the value of this asymptote from the graph (see Figure 3.1) or determine it algebraically as shown below.

$$C(t) = \frac{W(t)}{P(t)} \times 10^3 = \frac{14 + .67t}{154.31 + 2.40t} \times 10^3.$$

Divide both numerator and denominator of the fraction by t. The effect is to divide each term in the fraction by t, giving us

$$\frac{14 + .67t}{154.31 + 2.40t} \times 10^3 = \frac{\dfrac{14}{t} + .67}{\dfrac{154.31}{t} + 2.40} \times 10^3.$$

Now it is easier to see what happens for large values of t. As t gets larger and larger, both $\dfrac{14}{t}$ and $\dfrac{154.31}{t}$ get smaller and smaller. (We abbreviate this by writing $\dfrac{14}{t} \to 0$ and $\dfrac{154.31}{t} \to 0$.) Therefore

$$\frac{14 + .67t}{154.31 + 2.40t} \times 10^3 = \frac{\dfrac{14}{t} + .67}{\dfrac{154.31}{t} + 2.40} \times 10^3 \to \frac{.67}{2.40} \times 10^3 = 279,$$

and we can conclude that the horizontal asymptote is the line $y = 279$. This can be written in using "limit" notation as

$$\lim_{t \to \infty} C(t) = \lim_{t \to \infty} \frac{14 + .67t}{154.31 + 2.40t} \times 10^3 = 279.$$

This indicates that C(t) approaches the number 279 as t approaches infinity. This notation will be discussed in more detail in Section 3.5.

The graph of C(t) has as its horizontal asymptote the line y = 279. This means that if current trends continue then after a very long period of time, the US water withdrawal rate will stabilize at about 279 gallons per person per day. Remember, though, it isn't realistic to project too far into the future.

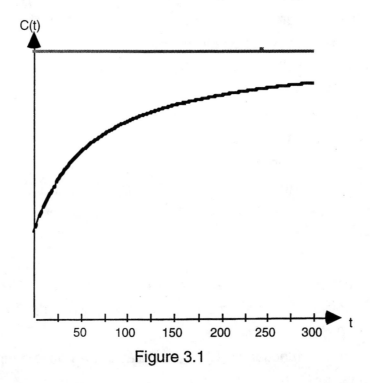

Figure 3.1

We can interpret the functions we have derived. If the trend shown by the 1950 - 1990 data continues, then

- population will continue to increase;
- water withdrawal for public supply will continue to increase;
- the per capita water withdrawal for public supply will continue to increase.

These statements are the respective interpretations of these properties of the functions:

- P(t) has positive slope;
- W(t) has positive slope;
- C(t) is an increasing function.

In the next section we see what the effect would be on water withdrawal if the per capita rate levels off or decreases. After that, we will look at total water withdrawal for all uses.

3.2   Change in Withdrawal Rates

In chapter two, we modeled two factors affecting water use, US population and daily water withdrawal for public use, with linear functions having positive slopes. The quotient of these two functions which described the per capita withdrawal was also increasing; that is, in any year the amount withdrawn per person is greater than the amount withdrawn the previous year. In this section we see first what would be the effect on water withdrawal if the per capita rate stayed at the 1997 level, then what would happen if the rate began to decrease in 1997. We still assume that the population will grow as described by the population function

$$P(t) = 2.40t + 154.31,$$

where $P(t)$ is the population in millions in year $1950+t$. Also, recall from the previous section (#1 in the group work) that in 1997, the per capita water withdrawal for public use was C(47) = 170 gallons per person per day.

Group Work

Suppose that the per capita withdrawal remains at the 1997 level of 170 gallons per person per day. This means that in any year 1997 or later, each person will use 170 gallons of water per day.

1.   Determine a new function $W_1(t)$ for withdrawal for public supply for $t \geq 47$. Use the same units as W(t).

2.   Give a verbal interpretation of the slope of $W_1(t)$.

3.   Graph $W_1(t)$ and W(t) on the same coordinate system.

4.   How much water would be withdrawn each day in 2010? How does this compare to your original prediction? (See Section 2.4, group work question #7.) How much water would be saved during that year if the withdrawal rate was held to 170 gallons per person per day?

5.   According to the function $W_1(t)$, the daily water withdrawal will still increase each year even though the per capita use stays constant. Why? What is the rate of this increase? How does this compare to the rate of increase determined in section 2.4 ?

★ ★ ★ ★

The original model (Section 3.1) can be used to predict the per capita withdrawal rate for any year. For example, in 1997 the predicted withdrawal is 170 gallons per person per day and if the trends continue, by 2010 each person will withdraw 182 gallons per day. (Why? Because 1997 - 1950 = 47 and C(47) = 170; 2010 - 1950 = 60 and C(60) = 182.) The next model (Group Work above) showed that even if per capita consumption did not increase at that rate but instead held constant at the 1997 level, there would still be an increase in the overall public withdrawal. Now we will see what would happen to the withdrawal rate if the per capita rate actually decreased.

Group Work

In the original model the daily per capita withdrawal rate (for public supply) increased from 170 gallons per person per day in 1997 to 182 gallons per person per day in 2010. Suppose that the trend is reversed and that between 1997 and 2010 the per capita withdrawal actually decreases from 170 to 160. We will assume that the per capita withdrawal falls at a constant rate, i.e., that the function is linear.

1.   How much must the daily per capita withdrawal decrease each year?

2.   Determine a new function to describe the per capita withdrawal between 1997 and 2010. (According to our assumption, this should be a linear function, and you already know how to write the equation for a linear function.) Call the function $C_2(t)$, where $C_2(t)$ is the per capita withdrawal in year 1950+t, measured in gallons per person per day.

3.   What is the domain of $C_2(t)$?

4.   Use the function $C_2(t)$ and the population function $P(t)$ to predict the daily withdrawal for public supply for the years 2000, 2005, and 2010.

5.   Determine a new function $W_2(t)$ to describe withdrawal for public supply for $t \geq 47$. In this case, in any year after 1997 you know the population, $P(t)$, and how much water each person withdraws daily in any year, $C_2(t)$. You want to combine these to determine $W_2(t)$, the amount of water the entire country will withdraw daily. The units are the same as $W(t)$. Here we use the new function

$C_2(t)$ from #1 above and the original $P(t)$ and we want to determine a new withdrawal function $W_2(t)$, $t \geq 47$.

* * * *

The function $W_2(t)$ is the product to two linear functions, that is, $W_2(t) = C_2(t) \times P(t)$. The result is a *quadratic* function (why?); its graph is a parabola. A quadratic function is a function of the form

$$f(x) = ax^2 + bx + c,$$

where *a, b,* and *c* are constants and $a \neq 0$. The graph of a quadratic function is a parabola which opens up if a > 0 and down if a < 0. The vertex of the parabola is the low point (or minimum) when the graph opens up and the high point (or maximum) when the graph opens down. The coordinate of the vertex has first coordinate $x = -\dfrac{b}{2a}$ and second coordinate $f\left(-\dfrac{b}{2a}\right)$. Explanations for these and other basic properties of quadratic functions are provided in Section 3.4.

Group Work, Continued

6.   Graph $W_2(t)$ and the original function $W(t)$ on the same coordinate system.

7.   How much water would be saved in 2010 if the new assumptions are valid?

8.   Determine the vertex of the parabola and give a verbal interpretation of the this information.

* * * *

The group work above shows that even if the daily per capita withdrawal decreases at a rate of almost a billion gallons each year, the total U.S. withdrawal for public supply will continue to increase for many years. This is because of the rate of growth for the population. At the end of the next section you will have a chance to look at other scenarios. You will try to find a rate of decrease for per capita withdrawal that will immediately decrease water withdrawal and will see what happens if the population growth slows.

3.3   Total Water Withdrawal

The pattern of total water withdrawal is less clear than domestic withdrawal. Table 3.1 below provides data of daily per capita withdrawal for all sources (Statistical Abstracts). This includes not only the personal use we looked at previously but also the water withdrawn for agriculture and industry.

Year	Daily Per Capita Withdrawal for all Uses (gallons per day)
1940	1027
1950	1185
1960	1500
1970	1815
1980	1953
1985	1650
1990	1620

Table 3.1

The Figure below is a scatter plot of the data on a (t,C) coordinate system, where t is the years since 1950 and C is the total per capita withdrawal, in gallons per day.

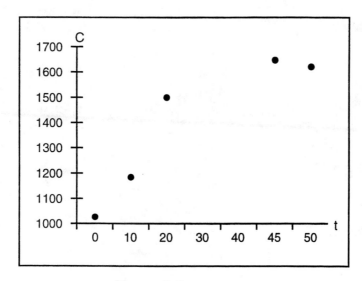

Figure 3.2

You can see that there was a fairly regular increase from 1940-1980, but then the pattern changed and the withdrawal actually decreased. Remember, this is per capita withdrawal, not total. The decrease in water withdrawal was primarily due to the fact that less water was used per person for irrigation and industrial purposed. In this section we will look at what this means for the total water withdrawal.

Group Work

We will use the most recent information on per capita withdrawal from 1985 to 1990, and assume that this trend will continue. Specifically, we make the following assumptions:

- *the per capita withdrawal decreased at a constant rate from 1650 gallons per person per day in 1985 to 1620 gallons per person per day in 1990,*
- *this trend continues;*
- *the population trend determined in section 2.3 continues.*

1. Determine a function which describes the daily per capita withdrawal in year 1950 + t. We continue to have t = 0 in 1950 since we will use this function together with the population function derived in section 2.3. However, since the assumptions are only for 1985 and later, that the domain of this function consists of all $t \geq 35$. Use the notation

TC(t) = Total per capita withdrawal in year 1950+t gallons per person per day.

2. Predict the per capita withdrawal in the year 2010.
3. Give a verbal interpretation of the slope.
4. Determine the function which describes the total daily water withdrawal ($\times 10^9$) gallons per day; denote it TW(t), where t = 0 in 1950.
5. According to your model, in what year will the total withdrawal be greatest? What is the predicted total withdrawal in that year? What is the per capita withdrawal? What is the population?

* * * *

3.3.1   Further Investigations

In the previous sections we examined patterns of water withdrawal under various assumptions. Here are some possibilities for additional investigations.

1. If each person in the United States cut personal water withdrawal gradually for the next ten years, is it possible to effect an immediate decrease in total domestic withdrawal? If so, how much would the per capita withdrawal rate need to be cut each year?

2.   Data shows that although water withdrawals for irrigation and industrial uses fell during the first part of the eighties, the amount used for irrigation stabilized while the amount used for industrial purposes actually rose in the last half of the decade. Even though the total amount increased, the per capita amount still declined due to the increasing population.   The total daily withdrawal actually increased from 403 billion gallons in 1985 to 411 billion gallons in 1990.

     a)   How much did the total daily withdrawal increase each year?

     b)   Assume that the total withdrawal continues to increase at this rate until at least 2010.   What would this say about the per capita daily withdrawal?

3.   Suppose that beginning in 1997 the population growth rate decreases by 10%. What is the new growth rate? Write a new population function using this growth rate which will be valid for $t \geq 47$. What would be the impact on water withdrawal, both for public supply and total withdrawal. Use the assumptions regarding per capita withdrawal made previously.

4.   Try to find the following information for your city or municipality. How much water per day does your local water system supply, and how many people use this water. If you can find both current and past information, you can try to model local withdrawals. If you use private well water, skip this question.

3.4   Quadratic Functions

A *quadratic function* has the form
$$f(x) = ax^2 + bx + c$$
where a, b and c are constants and a $\neq$ 0. Its domain consists of all real numbers.

The graph of a quadratic function is a *parabola* opening up if the leading coefficient a > 0 or opening down if a < 0. (See Figure 3.3.)

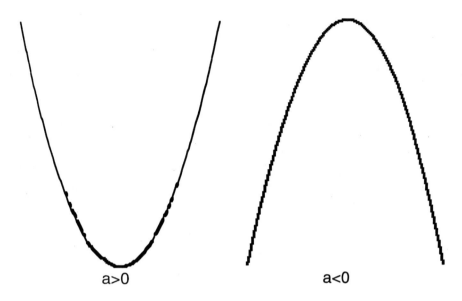

a>0   a<0

Figure 3.3

The graph of a quadratic function will always have one y-intercept and can have zero, one, or two x-intercepts (recall that intercepts are points where a graph crosses an axis). The y-intercept is the point where the graph crosses the y-axis so it will be the value of the function at x = 0. Since f(0) = c (c a constant), the y-intercept is the point (0,c). The x-intercept(s), if there are any, are the points where the graph crosses the x-axis. They can be found by replacing y with 0 and solving the quadratic equation

$$0 = ax^2 + bx + c.$$

This equation has one, two, or no real solutions. When it has one solution, there will be one x-intercept; when there are two solutions, there will be two x-intercepts; and when the equation has no real solution, the graph will have no x-intercepts. (See Figure 3.4.)

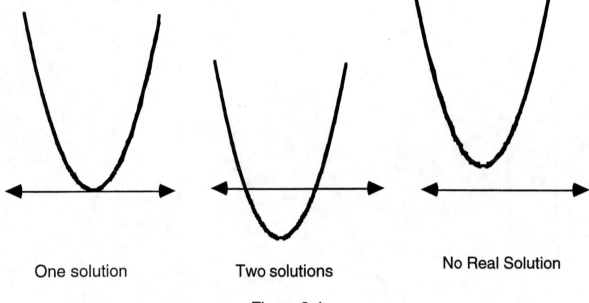

One solution   Two solutions   No Real Solution

Figure 3.4

3.4.1   Quadratic Equations and the Quadratic Formula

Any quadratic equation can be put into standard form

$$ax^2 + bx + c = 0$$

and solved using the *quadratic formula*. The derivation, which uses the method of completing the square, is relatively simple as the following steps illustrate.

First divide both sides by a,

$$x^2 + \frac{b}{a}x + \frac{c}{a} = 0,$$

then subtract $\frac{c}{a}$ from both sides to get

$$x^2 + \frac{b}{a}x = -\frac{c}{a}.$$

Next, add $\left(\frac{1}{2} \cdot \frac{b}{a}\right)^2$ to both sides,

$$x^2 + \frac{b}{a}x + \frac{b^2}{4a^2} = -\frac{c}{a} + \frac{b^2}{4a^2}$$

and simplify the right side and factor the left side

$$\left(x + \frac{b}{2a}\right)^2 = \frac{b^2 - 4ac}{4a^2}.$$

Take the square root of each side to get

$$x + \frac{b}{2a} = \frac{\pm\sqrt{b^2 - 4ac}}{2a},$$

and finally solve for x and simplify to get the famous quadratic formula,

$$x = \frac{-b \pm \sqrt{b^2 - 4ac}}{2a}.$$

The expression $b^2 - 4ac$ (called the *discriminant*) which is under the radical is significant. If $b^2 - 4ac > 0$, the quadratic equation has two real solutions; if $b^2 - 4ac = 0$, there is exactly one real solution; if $b^2 - 4ac < 0$, there is no real solution.

This formula can be used to find the solutions, if any, to a quadratic equation which is written in the standard form $ax^2 + bx + c = 0$.

Example 3.1

Solve $x^2 - 2x - 2 = 0$.

Solution

The coefficients are a = 1, b = -2 and c = -2. Then
$$\frac{-(-2) \pm \sqrt{(-2)^2 - 4 \cdot 1 \cdot (-2)}}{2 \cdot 1} = \frac{2 \pm \sqrt{12}}{2} = \frac{2 \pm 2\sqrt{3}}{2} = 1 \pm \sqrt{3};$$
the equation has two real solutions which are, accurate to two decimal places,

x = 2.73 or x = -.73.

* * * *

Example 3.2

Solve the equation $3x^2 - x + 3 = 1$.

Solution

First put the equation into standard form,
$$3x^2 - x + 2 = 0.$$
The coefficients are $a = 3, b = -1$ and $c = 2$. Then
$$\frac{1 \pm \sqrt{(-1)^2 - 4 \cdot 3 \cdot 2}}{2 \cdot 3} = \frac{1 \pm \sqrt{-8}}{6};$$
the equation has no real solutions since $b^2 - 4ac = -8 < 0$.

* * * *

Example 3.3

Find the x- and y- intercepts of the quadratic function
$f(x) = 1.5x^2 - 3.2x - 9.6$.

Solution

The y-intercept is found by replacing x with 0,
$f(0) = 1.5(0)^2 - 3.2(0) - 9.6 = -9.6$.

Thus the y-intercept is the point (0, -9.6). The x-intercepts are found by replacing y, or f(x), with 0 and solving for x,
$0 = 1.5x^2 - 3.2x - 9.6$.

Using the quadratic formula, we get

$$x = \frac{-(-3.2) \pm \sqrt{(-3.2)^2 - 4 \cdot 1.5 \cdot (-9.6)}}{2 \cdot 1.5}.$$

The solutions (rounded to two decimal places) are x = 3.81 and x = -1.68. The x-intercepts are (approximately) the points (3.81, 0) and (-1.68, 0). See Figure 3.5.

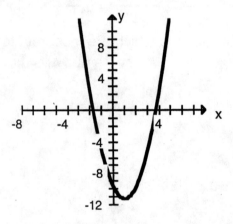

Figure 3.5

* * * *

The quadratic formula can be programmed into most graphing calculators. Read your manual for instructions.

Class Work

Decide whether each parabola opens up or down and determine the coordinates of the x- and y-intercepts.

1. $y = .02x^2 - 63.1x + 27$

2. $P(t) = -3t^2 + 27.5t + 13.1$

* * * *

3.4.2   Graphs of Quadratic Functions

A parabola which opens up has a lowest point, and one which opens down has a highest point. The highest or lowest point on a parabola is called the *vertex*. The parabola is symmetric about a vertical line through its vertex; this line is called the *axis of symmetry*. (See Figure 3.6.)

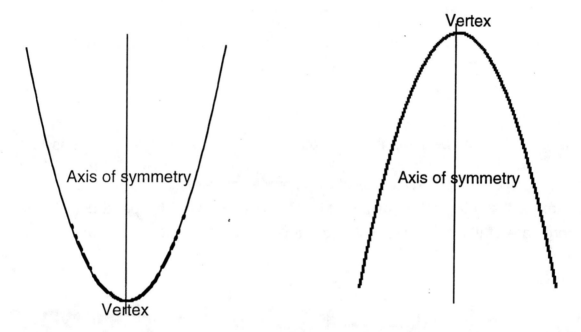

Figure 3.6

To find the coordinates of the vertex, we will look at an example. Consider the function $f(x) = 2x^2 - 8x + 7$ whose graph is shown in Figure 3.7.

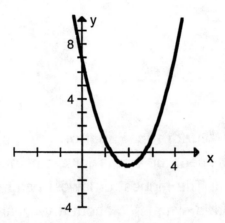

Figure 3.7

The axis of symmetry is the vertical line halfway between the two x-intercepts. These intercepts are given by the quadratic formula,

$$x = \frac{8 \pm \sqrt{8}}{4} = 2 \pm \frac{\sqrt{2}}{2} \ .$$

The point halfway between the intercepts $\left(2 - \frac{\sqrt{2}}{2}, 0\right)$ and $\left(2 + \frac{\sqrt{2}}{2}, 0\right)$ has first

coordinate x = 2, and so the axis of symmetry is the vertical line x = 2. Since the vertex is the intersection of the axis of symmetry with the parabola, the x-coordinate of the vertex is x = 2 and the y-coordinate is f(2) = -1. (See Figure 3.8.)

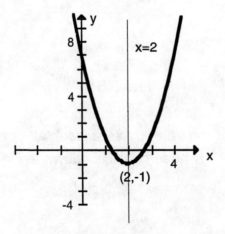

Figure 3.8

Note that the quadratic formula can be written in the form

$x = \dfrac{-b}{2a} \pm \dfrac{\sqrt{b^2 - 4ac}}{2a}$. The axis of symmetry is determined by the first term of this expression, i.e. $x = \dfrac{-b}{2a}$, and the coordinates of the vertex are $\left(\dfrac{-b}{2a}, f\left(\dfrac{-b}{2a}\right)\right)$.

These formulas hold even when there is no x-intercept.

Once we know the vertex of a parabola, we can determine the range of the quadratic function. The maximum or minimum value of the function will be the y-coordinate of the vertex. In the example above, $f(x) = 2x^2 - 8x + 7$ has a minimum value since the parabola opens up. The minimum value is -1 and it occurs when x = 2. Thus the range of the quadratic function is {y| y ≥ -1}.

In the next two examples, we will determine the intercepts, vertex, maximum or minimum and range of quadratic functions and use the information to graph each function.

Example 3.4

Graph the function $f(x) = -2x^2 + 12x + 13$; first determine its intercepts, vertex, maximum or minimum and range.

Solution

We compare $f(x) = -2x^2 + 12x + 13$ to the general quadratic function $f(x) = ax^2 + bx + c$ and set a = -2, b = 12, and c = 13. Since a < 0, the parabola opens down. The first coordinate of the vertex is $x = \dfrac{-12}{-4} = 3$, the second coordinate is $f(3) = 31$, and so the vertex is (3, 31). Thus the maximum value of the function is 31 and the range is {y|y ≤ 31}. The y-intercept is f(0) = 13 and the x-intercepts can be found by solving the equation 0 = -2x^2 + 12x + 13. Using the quadratic formula, the solutions (rounded to three decimal places) are x = 6.937 or x = -.937. Thus the x-intercepts are the points (6.937, 0) and (-.937, 0). The graph is shown in Figure 3.9.

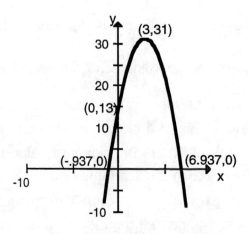

Figure 3.9

* * * *

Example 3.5

Graph $f(x) = 0.4x^2 - x + 7.3$; determine the intercepts and vertex.

Solution

The parabola opens up since a = 0.4 is positive. The vertex has coordinates $x = \dfrac{1}{.8} = 1.25$ and $y = f(1.25) = 6.675$. The function has a minimum value of 6.675 and its range is {y| y ≥ 6.675}. The y-intercept is f(0) = 7.3 and the x-intercepts are found by solving the equation $0 = 0.4x^2 - x + 7.3$.

Using the quadratic formula,

$$x = \frac{-(-1) \pm \sqrt{1 - 1168}}{.8} = \frac{1 \pm \sqrt{-1068}}{.8}.$$

This equation has no real solution since the quantity under the radical is -10.68, a negative number. Thus the graph has no x-intercepts and does not cross the x-axis. (We should have known that there were no x-intercepts when we found that the minimum was 6.675.). The graph is shown Figure 3.10.

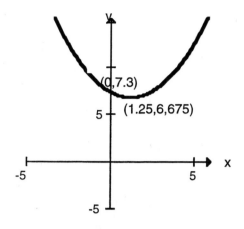

Figure 3.10

* * * *

Class Work
3.   Find the vertex of the parabola $b(z) = 5.6z^2 - 4.3z + 8.1$.

4.   Determine the maximum or minimum value and the range of the quadratic function $H(x) = 2.1x^2 + x - 15$.

* * * *

In the next two examples we will look at solutions to some quadratic equations and what they mean relative to a graph.

Example 3.6
Solve the equation $x^2 - 2.4x + 144 = 15$.

Solution
Figure 3.11 shows the graph of the parabola
$$y = x^2 - 2.4x + 144$$
together with the horizontal line
$$y = 15.$$
The solutions to the equation $x^2 - 2.4x + 144 = 15$ are the x-coordinates of the points where the line y = 15 and the parabola intersect. Use the calculator or computer to estimate these values to be x = 5.07 or x = -2.67 (rounded to two places).

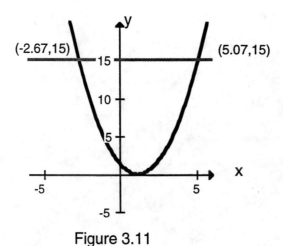

Figure 3.11

It is, of course, possible to solve this equation algebraically. First write the equation in standard form by subtracting 15 from both sides,

x² - 2.4x - 13.56 = 0;   (note: you effectively raise the x-axis.)

then use the quadratic formula to get solution x = 5.07 or x = -2.67.

** * * ***

Example 3.7

Solve the equation $5 = .4x^2 - x + 7.3$.

Solution

The graph of $y = .4x^2 - x + 7.3$ is shown in Figure 3.10 (Example 5 above). The minimum value of the parabola is 6.675 and hence the parabola cannot possibly intersect the line y = 5. (Note that if we first write the equation in standard form $0 = .4x^2 - x + 2.3$ then solve using the quadratic formula, we see that the equation has no solution.)

** * * ***

Class Work

Solve each of the following equations and indicate the solutions (if any) on a graph. If the equation has no solution use a graph to show why.

1. $5 = x^2 - 6x - 10$

2. $2.5x^2 - x + 7.3 = -1$

** * * ***

Exercises

Graph each quadratic function by determining whether its graph opens up or down and by finding its vertical intercept, its horizontal intercepts (if any) and its vertex.

1. $f(x) = -2.1x^2 + 3.5x + 1.7$
2. $f(x) = 5.04x^2 + 11.21x + 41.68$
3. $l(t) = .67t^2 - 1.33t$
4. $h(s) = -3s^2 - 8$
5. $s(t) = .0006t^2 - .126t + 12.9$

* * * *

3.5   Rational Functions

A *rational function* is a quotient of two polynomial functions. The domain consists of all real numbers for which the denominator is not zero. (Of course, the denominator of a rational function cannot be the zero polynomial.) In the following examples we look at some graphs and properties of rational functions.

Example 3.8

Graph $f(x) = \dfrac{1}{x}$.

Solution

Figure 3.12 shows the graph of $y = \dfrac{1}{x}$.

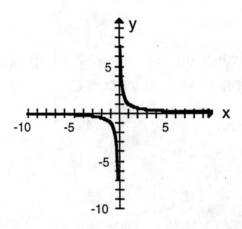

Figure 3.12

The graph does not cross the y-axis since 0 is not in the domain of the function and does not cross the x-axis since there is no solution to the equation $\frac{1}{x} = 0$.

Also notice:

   1)   as x gets larger and larger, y gets smaller and smaller, i.e., close to 0 (we write as $x \to \infty$, $y \to 0$);

   2)   as $x \to -\infty$ (i.e. as Ixl is larger and larger but x < 0) then $y \to 0$;

   3)   as $x \to 0^+$ ("x approaches 0 from the positive side"), $y \to \infty$;

   4)   as $x \to 0^-$ ("x approaches 0 from the negative side"), $y \to -\infty$.

It is quite evident that the graph of $y = \frac{1}{x}$ gets closer and closer to the x-axis as $x \to \pm\infty$ and closer and closer to the y-axis as $x \to 0^{\pm}$ (from either side.) Because of these properties the coordinate axes are called *asymptotes*. The line x = 0 is called a *vertical asymptote* and the line y = 0 is a *horizontal asymptote*.

★ ★ ★ ★

Limit Notation

   Limit notation provides a compact way of expressing the concept of taking the limit of a function. Although a complete discussion of taking the limit of a function is beyond the scope of this text, understanding the meaning of the notation is useful. In Example 3.8 above we were presented with four cases when x approached a certain value. For each of the cases we will write the notation and the result.

$\lim\limits_{x \to \infty} \dfrac{1}{x} = 0$, which says that as x approaches positive infinity, y approaches

zero.

$\lim\limits_{x \to -\infty} \dfrac{1}{x} = 0$, which says that as x approaches negative infinity, y

approaches zero.

$\lim\limits_{x \to 0^+} \dfrac{1}{x} = +\infty$, which says that as x approaches zero from the positive

(right) side, y approaches positive infinity.

$\lim\limits_{x \to 0^-} \dfrac{1}{x} = -\infty$, which says that as x approaches zero from the negative (left)

side, y approaches negative infinity.

$$* \quad * \quad * \quad *$$

Example 3.9

Graph $R(x) = \dfrac{1}{x-2}$ and determine its asymptotes.

Solution

Use your calculator or computer to see the graph as shown in Figure 3.13.

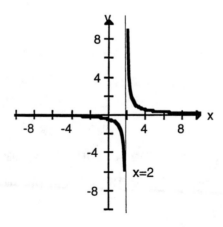

Figure 3.13

The domain of R(x) consists of all real numbers x ≠ 2, and so the graph will never cross the vertical line x = 2. However, using our notation from the previous example, observe that

as $x \to 2^-$ then $R(x) \to -\infty$

and

as $x \to 2^+$ then $R(x) \to \infty$

The line x = 2 is a vertical asymptote for the graph.

The equation $\dfrac{1}{x-2} = 0$ has no solution so the graph will not cross the x-axis (that is, it does not have an x-intercept.). You can see from the graph that the range consists all real numbers y ≠ 0. Moreover,

as $x \to \infty, R(x) \to 0$,

and

as $x \to = -\infty, R(x) \to 0$,

and the line y = 0 is a horizontal asymptote.

Notice that the graph of R(x) is just the graph of f(x) shifted two units to the right. This graph does have a y-intercept, (0,-2), but no x-intercept.

* * * *

The line y = c is a *horizontal asymptote* for the graph of the function f(x) if $f(x) \to c$ as $x \to \pm\infty$. The line x = a is a *vertical asymptote* for the graph of the function f(x) if $|f(x)| \to \infty$ as $x \to a$. Graphs can have other types of asymptotes, but we won't be concerned with these at this time.

Example 3.10

Find the vertical and horizontal asymptotes of $F(x) = \dfrac{x-2}{x+3}$. Sketch the graph.

Solution

This one is different from the first function since the numerator is not constant. In this case we divide the denominator into the numerator,

$$\frac{x-2}{x+3} = 1 - \frac{5}{x+3}.$$

In this form it is easy to determine the asymptotes. The only number not in the domain is x = -3 so the graph does not cross the vertical line x = -3. We can see that as x approaches -3 from the left,

$$x \to -3^-, \frac{5}{x+3} \to -\infty$$

and so

$$F(x)=1-\frac{5}{x+3}\to\infty;$$

and as x approaches -3 from the right,

$$x\to-3^{+},\frac{5}{x+3}\to\infty$$

and so

$$F(x)=1-\frac{5}{x+3}\to-\infty.$$

Using the limit notation from above we write

$$\lim_{x\to-3^{-}}F(x)=\infty \text{ and } \lim_{x\to-3^{+}}F(x)=-\infty;$$

the line x = -3 is a vertical asymptote.

To find the horizontal asymptote, note that as

$$x\to\pm\infty,\frac{5}{x+3}\to0$$

and so

$$F(x)=1-\frac{5}{x+3}\to1.$$

Again, we can use the limit notation and write

$$\lim_{x\to\pm\infty}F(x)=1$$

Thus the horizontal asymptote is y = 1. The graph is shown in Figure 3.14.

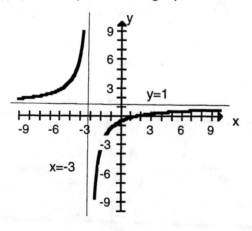

Figure 3.14

* * * *

In general, a rational function, $R(x)=\frac{p(x)}{q(x)}$, will have a vertical asymptote

x = c if c is a zero of the denominator q(x).

83

Example 3.11

Find the vertical and horizontal asymptote(s) of $R(x) = \dfrac{x}{x^2 - 1}$. Sketch the graph.

Solution

First factor the denominator
$$x^2 - 1 = (x - 1)(x + 1).$$
Thus the lines $x = -1$ and $x = 1$ are vertical asymptotes.

Finding the horizontal asymptote requires a little more work. We must determine how the function behaves as $x \to \infty$ and as $x \to -\infty$. This is easy if we first divide both numerator and denominator of the rational function by the highest power of the variable x in the expression, in this case by x^2,

$$R(x) = \frac{x}{x^2 - 1} = \frac{\dfrac{x}{x^2}}{\dfrac{x^2 - 1}{x^2}} = \frac{\dfrac{1}{x}}{1 - \dfrac{1}{x^2}};$$

as $x \to \pm\infty$, $\dfrac{1}{x} \to 0$ and $\dfrac{1}{x^2} \to 0$.

Therefore $R(x) \to \dfrac{0}{1 - 0} = 0$ and the horizontal asymptote is y = 0. The graph is shown in Figure 3.15.

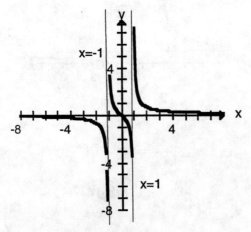

Figure 3.15

* * * *

84

Exercises

1.   Use the graphs shown to find:

   (i) the domain and range of each function;

   (ii) the intercepts, if any;

   (iii) vertical asymptotes, if any;

and

   (iv) horizontal asymptotes, if any.

a.

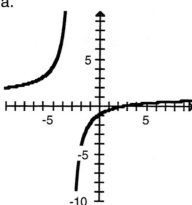

b.

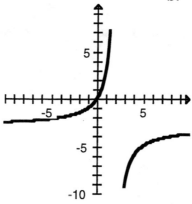

c.

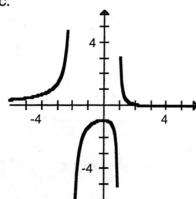

2.   Refer to the graph shown below. Determine the equations of the vertical and horizontal asymptotes and write statements describing the behavior of the function near each of the asymptotes.

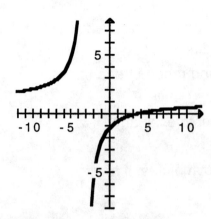

3.   Graph each rational function. Determine the domain, range and asymptotes.

a. $f(x) = \dfrac{2x-3}{x+2}$

b. $g(x) = \dfrac{x+2}{x^2-4}$

c. $h(x) = \dfrac{4-5x}{2x-3}$

* * * *

CHAPTER FOUR
CONSTANT FLOW RIVERS

Introduction

Water allocation is a very important and often controversial issue. In the early days of this country, during the "discovering" of the American West, ranchers and farmers would often dam streams and rivers so that they could have enough water for their cattle and for irrigation of their crops. This would understandably anger other ranchers and native peoples who lived downstream and were also dependent upon the stream for their water supply. Violent fights over water rights occurred in many instances like this, so government regulation of water allocation became necessary. However, even today many legal battles are being fought over water usage, and as our population increases so does the number of such cases. Industry requires more and more water to produce consumer products, more is needed for irrigation of crops and watering of livestock for food production, and of course, for human consumption.

The issue of water allocation is more crucial in the southwestern United States than in other parts of the country. Indeed, the Colorado River which, together with all its tributaries supplies almost all of the southwest with water, never reaches its natural destination, the Pacific Ocean. It's as if the human population served by the Colorado basin were to increase by just one person, then everyone else would get just a little less water.

Easterners enjoy much more water affluence than westerners. Many southwestern rivers are either dry or run at very low volume during certain times of the year whereas most eastern rivers, during a normal year, carry a fairly constant volume if not regulated by a dam. But even in the east, court battles regarding rights to river water are springing up more and more as demands grow.

In order to determine water allocations, one piece of information is obviously necessary: the amount of water available. For rivers and streams, the measure of available water is streamflow. *Streamflow* is the volume of water which passes through a cross-section of a river in a specified interval of time. In the U. S. streamflow is usually measured in cubic feet per second (*cfs*). If the streamflow for Turkey Creek at the highway bridge is 25 *cfs*, this means that 25 cubic feet of water in Turkey Creek runs under the bridge each second. Other units of measure can also be used, for example, cubic meters per minute, or cubic feet per month, but the measurement of streamflow must always be volume per unit of time. Streamflow will clearly vary at different times of the year, at different locations on the stream, and from year to year. Because of these variations and diverse demands for water, the prediction of streamflow is a very important business. A major factor in this prediction is the amount of precipitation over the stream's watershed. Precipitation amounts have been recorded for years in many regions of the U. S. and average precipitation data can be used along with other factors to predict streamflow. Records of streamflow have also been kept; these data are gathered by actual measurement at various points along the stream.

In this chapter we will look at a simplistic method of predicting streamflow along a small river with an assumed constant flow. This analysis of streamflow will use certain algebraic functions known as *polynomial functions*. The polynomial functions needed for this study will be introduced in the discussion, then will be studied in more generality and depth in later sections of this chapter.

4.1 Predicting streamflow for Chinle River

The Chinle River has its headwaters in a warm climate and derives all its water from precipitation in the region and is unaffected by snow or ice. The precipitation is fairly constant throughout the year and most of this precipitation finds its way into the ground water feeding the river. Since the Chinle River acquires most of its flow from ground water its flow is nearly constant. A coal-burning power plant operates on the river and draws its water from the river for cooling purposes. Just below the village, Chinle River flows into a lake which is a popular fishing and boating place. In order to support these recreations, it is necessary that the lake be kept at a certain level. It is important to know the amount of water that Chinle River can supply so that a fair balance among these requirements can be maintained.

We begin by predicting streamflow at any point on the river from its origin to its entrance into the lake. The prediction is modeled primarily upon the regional precipitation, observed ground water flow and the size of the watershed; the relevant information is listed below (these are our assumptions).

i) Annual precipitation is 4 feet uniformly distributed throughout the year.

ii)   The watershed from the origin of the river to the lake is shaped roughly like a triangle with one vertex at the headwaters. The river flows in a straight line through the triangle for a distance of 100 miles and empties into the lake at the side opposite its origin; the distances from the entrance of the river into the lake to each of the other vertices are, respectively 50 miles and 30 miles. (See Figure 4.1.) The geology of the watershed is such that all water flows to the river and out a narrow short valley to the lake.

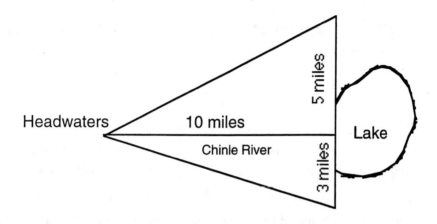

Figure 4.1

iii)   The drainage in the watershed is in a direction perpendicular to the river bed; thus the watershed for a section of the river extending from its origin to a point x miles downstream will be the sub-triangle which is the corresponding portion of the entire watershed. (See Figure 4.2)

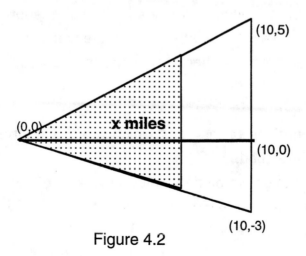

Figure 4.2

iv) Not all of the precipitation which falls on the watershed actually reaches Chinle River; much of it is taken by evaporation and evapotranspiration. As the distance downstream increases, the percentage of precipitation from the corresponding watershed which reaches the river bed decreases. It is estimated that this percentage varies linearly from 60% at the headwaters to 20% at the lake entrance. For instance, at fifty miles from the headwaters 40% of the precipitation in the area from the headwaters to five miles from the lake will reach the lake.

Based on the above assumptions, we can now predict the streamflow for Chinle River at any point.

4.1.1 Area of the watershed

Let x denote the number of miles downstream from the origin of the river. Streamflow is determined by the amount of precipitation over the watershed which actually enters the river bed. Therefore the first thing which must be determined is the total volume of water which falls over the watershed for any portion of the river from 0 miles to x miles. We know the depth of the rainfall; the volume of water is the depth multiplied by the area of the watershed, so we must determine the area of the watershed. Refer to Figure 4.2. We place the watershed for Chinle River on a coordinate system with the headwaters at the origin and with the river flowing along the positive x-axis over the interval [0, 10]. Thus the entrance to the lake is at the point (10, 0), and the watershed extends vertically 5 miles up and 3 miles down from this point. The watershed for the portion of the river from 0 miles to x miles is the sub-triangle shown in Figure 4.2. Follow the steps below to find the area of the watershed for the portion of the river from 0 miles to x miles, then answer the related questions.

Group Work
1.   Write the equations of the lines which are the northern and southern boundaries of the watershed.
2.   Use these equations to determine the area A(x) of the watershed for the river from 0 to x miles.
3.   What is the area of the watershed for the river from 0 to 3 miles?
4.   What is the area of the watershed for the river from its origin to the lake?

* * * *

4.1.2 Percentage of precipitation which enters the river bed

We must determine the percentage of water which contributes to the streamflow for Chinle River. Recall that this decreases linearly from 60% at the origin to 20% at the lake.

Group Work

1.   Determine the function which gives the percentage $D(x)$ of water from precipitation which reaches the portion of the river from 0 to x miles.

2.   What percentage of precipitation reaches the river at 7.2 miles?

* * * *

4.1.3 Total volume of precipitation over the watershed

Recall that the volume of water from precipitation is the depth multiplied by the area of the watershed. We must be careful about the units of measurement when determining this. Since we will measure streamflow in cubic feet per second, it will be convenient to measure the volume of precipitation in cubic feet.

Group Work

1.   Determine the function $V(x)$ which gives the annual volume of precipitation in cubic feet which falls over the watershed for the river from 0 miles to x miles. Note that the amount of rainfall is already given in feet but the area of the watershed as determined by the function $A(x)$ is in square miles. Therefore you must first convert this area to square feet.

2.   What is the volume of precipitation which falls over the watershed for the portion of the river from 0 to 5.1 miles?

3.   What is the volume of precipitation which falls over the entire watershed for the river?

4.   Graph the function $V(x)$.

* * * *

4.1.4   Streamflow

Now we are ready to achieve our goal, i.e., to determine the streamflow for Chinle River. Recall that the streamflow is predicted by the amount of water which actually reaches the river bed.

Group Work

1. Use the percentage function D(x) and the volume function V(x) to determine the function F(x) which gives streamflow for the river at any point x miles downstream for its origin. The measure for streamflow should be cubic feet per second; the volume V(x) is in cubic feet per year so you must first convert this to ft^3/sec. Round coefficients to two decimal places.

2.   What is the predicted streamflow for Chinle River at 2.2 miles?

3.   What is the predicted streamflow for Chinle River at the entrance to the lake?

4.   What is the amount of water which flows from Chinle River into the lake each year?

5.   Graph the function F(x).

(Note: an optional section at the end of this chapter examines the effect of a coal burning power plant on the waters of Chinle River.)

* * * *

4.2   Graphs Of Polynomials Functions

A *polynomial function* has the form
$$f(x) = a_0 + a_1 x + a_2 x^2 + ... + a_n x^n$$
where $a_0, a_1, a_2, ..., a_n$ are constants, n is a non-negative integer and $a_n \neq 0$. The largest exponent, n, is the degree of the polynomial. The domain consists of all real numbers; the range will be considered later.

Example 4.1

Determine whether each of the following functions is a polynomial. For those that are, give the degree; for those that are not, explain why not.

a. $f(x) = 5 - 3x^3$

b. $g(x) = -5$

c. $F(x) = \dfrac{1}{x}$

d. $G(x) = \sqrt{x}$

Solution

a.   f is a polynomial function of degree 3.

b.   g is a polynomial function of degree 0.

c.   F is not a polynomial function since $\dfrac{1}{x} = x^{-1}$.

d.   G is not a polynomial function. It contains a square root of x, which is x raised to the $\dfrac{1}{2}$ power.

★ ★ ★ ★

You are already familiar with some polynomial functions.

*   A *constant function*, $f(x) = c$, is a polynomial function of degree 0; its graph is a horizontal line.

*   A *linear function*, $f(x) = ax + b$, is a polynomial function of degree 1 whose graph is a line with slope a.

*   A *quadratic function*, $f(x) = ax^2 + bx + c$, is a polynomial function of degree 2 whose graph is a parabola.

In this section we will look at the graphs of polynomial functions of degrees 3, 4, 5 and 6. We begin with the simplest such polynomials, functions of the form

$$f(x) = x^n, \, n \le 6.$$

(Polynomials with only one term are called *monomials*.)

We will first look at functions $f(x) = x^n$ where n is an even integer.

Figure 4.3 below shows the graphs of y = x², y = x⁴, and y = x⁶.

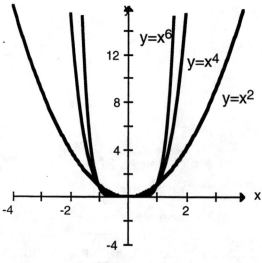

Figure 4.3

You can see that the graphs of y = x⁴ and y = x⁶ look very much like the parabola y = x², but they are not parabolas. As the power increases, the graph of y = xⁿ becomes flatter for -1 < x < 1 and more nearly vertical for very large positive and negative values of x. The range of y = xⁿ consists of all non-negative real numbers since xⁿ ≥ 0 for all real numbers x if n is an even integer.

Next we look at the function $f(x) = x^n$ where n is an odd integer greater than 1. Figure 4.4 below shows the graphs of y = x³ and y = x⁵.

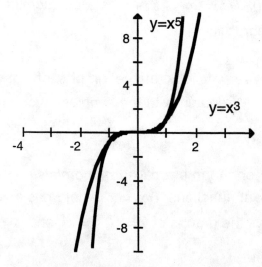

Figure 4.4

These graphs have common properties:

    i)   the point (0,0) is the only intercept (both x- and y-intercept);

    ii) the range consists of all real numbers.

These properties are characteristic of the graph of $y = x^n$ for any odd integer n. Also, as the power n increases the graph becomes flatter when x is between -1 and 1 and steeper outside of this interval. Notice the behavior of these functions for values of x far from zero: as x takes on negative values of larger magnitude (written $x \rightarrow -\infty$) the graph of $y = x^n$ falls without bound; as x takes on larger and larger positive values (written $x \rightarrow \infty$) the graph rises without bound.

Class Work

    Use your graphing calculator or computer software package to graph $y = x^4$ and $y = x^6$ on the same screen. Then zoom in on the portion of the graph where -1 < x < 1. Which graph is closer to the x-axis on this interval? Repeat this for $y = x^3$ and $y = x^5$.

* * * *

    Next we will look at graphs of the form $f(x) = ax^n$ where n is any positive integer ≤ 6, and a is a non-zero real number. Follow through the steps outlined below to analyze these graphs.

Class Work

1.   Graph $y = x^3$ and $y = 3x^3$ on the same screen. What do you observe?

2.   Now graph $y = x^3$ and $y = (1/3)x^3$ on the same screen. What do you observe?

3.   Next graph $y = x^3$ and $y = -x^3$ on the same screen. Again state your observations.

4.   Can you generalize the results to any function of the form $f(x) = ax^n$?

* * * *

    The graph of $f(x) = ax^n$ will have one of the four shapes shown in Figure 4.5 depending on the sign of a and whether the exponent n is even or odd.

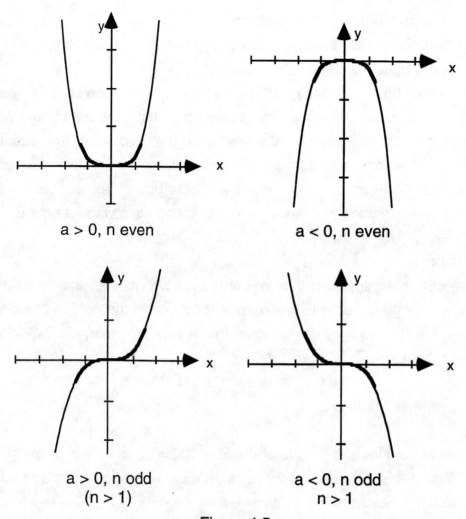

a > 0, n even   a < 0, n even

a > 0, n odd   a < 0, n odd
(n > 1)   n > 1

Figure 4.5

In the next examples, observe how the basic graphs are shifted.

Example 4.2

   Graph g(x) = x^3 - 2. Label the intercepts and determine the domain and range.

Solution

   The graph is shown in Figure 4.6

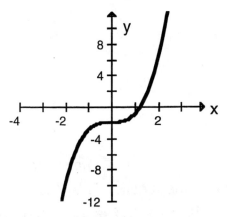

Figure 4.6

Notice that this graph looks like the graph of f(x) = x^3 shifted down two units. However, the domain and range of both of them consist of all real numbers. The y-intercept of the graph of y = x^3, (0,0), is shifted down two units to (0,-2). There will be a single x-intercept (1.26,0) which can be estimated from your graph by tracing to the point where the graph crosses the x-axis. It is also simple to find this x-intercept algebraically. Solve the equation

$$x^3 - 2 = 0$$
$$x^3 = 2$$
$$x = \sqrt[3]{2} = 1.26 \text{ (rounded)},$$

so we estimate the x-intercept is (1.26, 0).

＊ ＊ ＊ ＊

Example 4.3

   Graph h(x) = -x^4 + 1. Label the intercepts and give the domain and range.

Solution

The graph of h(x), shown in Figure 4.7, looks like the graph of
$y = -x^4$ shifted up 1 unit.

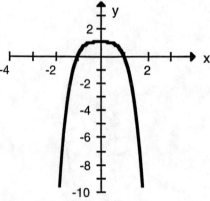

Figure 4.7

The y-intercept is (0,1) (since h(0) = -0⁴ + 1) and the x-intercepts are (-1,0) and
(1,0). The maximum value of the function is y = 1. The domain consists of all
real numbers and the range is the set of real numbers less than or equal to 1.
Again the x-intercepts can also be determined algebraically; solve

$$-x^4 + 1 = 0$$
$$1 = x^4$$
$$\pm 1 = x.$$

Thus the x-intercepts are (-1,0) and (1,0).

* * * *

The graph shown in the next example is a bit different.

Example 4.4

Graph the function F(x) = (x-3)(x-1)(x+2) on your calculator. Give the
intercepts and determine the domain and range.

Solution

Figure 4.8 shows the graph.

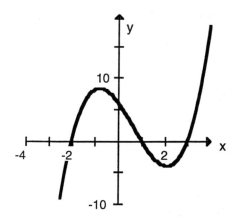

Figure 4.8

The y-intercept is (0,6); the x-intercepts are the points (3,0), (1,0) and (-2,0). The domain and range consist of all real numbers.

Of course, we could have determined the intercepts without the graph. The y-intercept is (0,6) since f(0) = 6; to determine the x-intercepts, replace f(x) with 0 and solve

0 = (x-3)(x-1)(x+2).

Setting each factor equal to zero we get

x-3 = 0, x-1 = 0 and x+2 = 0,

and thus the solutions are x = 3, x = 1 and x = -2.

This function has neither a maximum (largest value) nor minimum (smallest value) but the graph does have one "local" high point and one low point. Using the trace feature on the calculator, we can estimate the coordinates of these points, the high point is (-.79, 8.21) and the low point is (2.12, -4.06). These points are called relative maximum and relative minimum points. We say that the function has a relative maximum of 8.21, which occurs when

x = -.79,

and a relative minimum of -4.06 which occurs when

x = 2.12.

★ ★ ★ ★

If a polynomial function is written in factored form, it is easy to determine the x-intercepts. However, most polynomial functions aren't in factored form and polynomials of degree 3 and higher can be difficult or even impossible to factor. Thus you will have to estimate the x-intercepts on the graph. For practical purposes estimates that are "close enough" can be obtained by zooming in on the intercepts using a graphing calculator, using the "solve" command or "zero" command on the calculator, or by using a computer plotting program for polynomials of degree 3 or higher.

Example 4.5

Graph the function $f(x) = -x^3 + 7x^2 + 36x - 240$ on your calculator. Label the intercepts and estimate the coordinates of the relative maximum and minimum points.

Solution

Figure 4.10 shows the graph on a viewing screen with
$-10 \le x \le 10$ and $-10 \le y \le 10$.

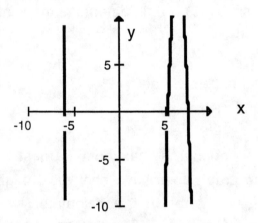

Figure 4.9

This doesn't show many of the significant features of the graph. We can see that there are at least three x-intercepts but the top and bottom of the graph are cut off, and we need to determine a reasonable viewing rectangle. The y-intercept is (0,-240) so we should choose a number less than -240 for the minimum y value. Figure 4.10 which shows the graph with $-10 \le x \le 10$ and $-250 \le y < 50$ gives a better picture but the low point is still cut-off.

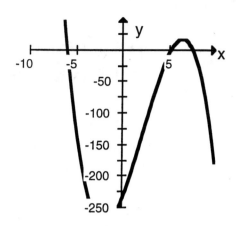

Figure 4.10

We change the minimum y-value to -300 (Figure 4.11) and can now see all of the significant points on the graph.

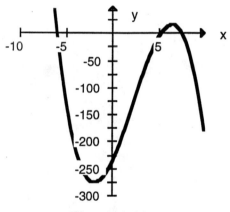

Figure 4.11

We estimate the x-intercepts to be (7.56,0), (5.36,0) and (-5.92,0). The relative maximum point is approximately (6.51, 15.13) and the relative minimum point is approximately
(-1.84, -276.31).

* * * *

Example 4.6

Graph $f(x) = x^4 - 2x^3 + 1$. Label the intercepts and determine the coordinates of any relative maximum or minimum points.

Solution

Since f(0) = 1, the y-intercept is (0,1). This is not very large so we might try a viewing screen with -5 ≤ x ≤ 5 and -3 ≤ y ≤ 5 (see Figure 4.12). We see that there are two x-intercepts: (1,0) and (1.84, 0). The function has one relative minimum point, (1.5, -.6875) which is, in fact, the lowest point on the graph. Thus the absolute minimum value of the function is -.6875 and the range consists of all real numbers y greater than or equal to -.6875.

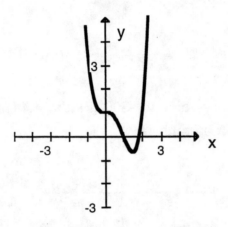

Figure 4.12

* * * *

Example 4.7

Graph f(x) = -2x^4 + x^3 + 13 x^2 + 8x -1. Give the coordinates of all intercepts, relative maximum and minimum points and determine the range.

Solution

The y-intercept is (0,-1); we will estimate the x- intercepts. With a little "trial and error", you should be able to determine a viewing screen that will show all of the significant points on the graph. Figure 4.13 shows the graph with the viewing screen with -4 ≤ x ≤ 4 and -10 ≤ y ≤ 50.

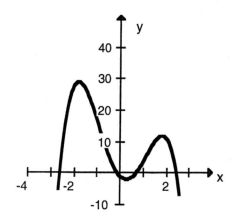

Figure 4.13

You can see that there are four x-intercepts: (-1.80,0), (-.86, 0), (.11, 0) and (3.05, 0). There are two relative maximum points, (2.13, 43.52) and (-1.42, 2.86); the second of these is also the absolute maximum point. There is one relative minimum point,

(-.33, -2.28). The range consists of all real numbers y less than or equal to 43.52.

* * * *

If you study calculus, you will learn that the graph of every polynomial function is a smooth, continuous curve. It will contain no sharp corners or breaks in the graph. In the next section, we will learn examine the number of roots of polynomial function of a given degree.

Exercises
Graph each of the following polynomial functions. Give the coordinates of all intercepts, relative maximum and minimum points and state the range of the function.

1. $f(x) = x^3 + 8$
2. $g(x) = 2x^3 + x - 3$
3. $p(x) = -3x(x+2)(x-1)$
4. $s(x) = 16 - x^4$
5. $h(x) = x^3 + x^2 - 5x - 2$
6. $q(x) = x^4 + x^3 - 1$
7. $r(x) = -2(x + 1)^2(x^2 + 1)$

8. $F(x) = -2x^3 + 3x^2 + x + 2$
9. $H(x) = x^4 - 2x^3 + 1$
10. $G(x) = x^4 + 1.1x^3 - x^2 + 0.9x - 3.2$
11. Which of the following statements are true about the graph of the polynomial function $f(x) = x^3 + bx^2 + cx + d$? Explain your answers.
    a. It intersects the y-axis exactly once.
    b. It might not intersect the x-axis.

* * * *

4.3 Roots of Polynomial Functions, Factor and Remainder Theorems

For any polynomial function f, if c is a number for which $f(c) = 0$, then c is a *root* or *zero* of $f(x)$. Thus the real roots or zeros of a polynomial function are the x-intercepts. When graphing a polynomial function it is helpful to know how many roots there are and to estimate their values. We already know that a polynomial function of degree 1, a linear function, has one root and a polynomial function of degree 2, a quadratic function can have 0, 1, or 2 roots. The graphs of third and fourth degree polynomials in the previous set of exercises suggest that a third degree polynomial can have at most three roots and a fourth degree polynomial can have no more than four. This is indeed true as we will soon see.

It is easy to determine the roots of a polynomial function if it is given in factored form; simply replace y, or $f(x)$, with zero and solve for x.

Example 4.8

Find all of the roots of each polynomial function.
a. $f(x) = (x+1)(x-2)(x+4)$
b. $t(x) = -2(x+1.5)(x-3)(x-5)^2$
c. $g(x) = (x^2+1)^2$

Solution

a. Replacing $f(x)$ with zero, we get
$$0 = (x+1)(x-2)(x+4)(x-2)^2$$
which gives us
$$x + 1 = 0, \text{ or } (x - 2)^2 = 0$$

$$x = -1 \text{ or } x = 2.$$

There are two roots, -1 and 2.

b.   Replacing t(x) with zero, we get

$$0 = -2(x + 1.5)(x - 3)(x - 5)$$

which gives us

$$x + 1.5 = 0 \text{ or } x - 3 = 0 \text{ or } x - 5 = 0$$

$$x = -1.5 \text{ or } x = 3 \text{ or } x = 5.$$

There are three roots, -1.5, 3 and 5.

c.   Replacing g(x) with zero, we get

$$0 = (x^2 + 1)^2$$

which has no real solutions since $x^2 + 1$ is never zero. Therefore there are no real roots.

$$* \quad * \quad * \quad *$$

There is a simple relationship between factors of polynomials and roots of polynomial functions. Any polynomial f(x) can be divided by any non-zero polynomial g(x), giving a quotient q(x) and remainder r(x), where the degree of r(x) is less than the degree of the divisor g(x),

$$f(x) = g(x)q(x) + r(x).$$

This is known as the *division algorithm*. If g(x) is a factor of f(x) then the remainder r(x) = 0, and conversely, whenever the remainder is zero, g(x) is a factor of f(x). In particular, any polynomial f(x) can be divided by a first degree polynomial of the form g(x) = x - c, c a real number, to get

$$f(x) = (x - c)q(x) + r, \text{ where r is constant.}$$

Thus, for any real number c, x - c is a factor of f(x) if and only if the remainder r = 0. This gives us a way to determine if a number c is a root of f(x) without actually carrying out the division. This can be deduced from the following theorem.

Remainder Theorem   If a polynomial function is divided by x - c, then the remainder is f(c).

To see that the Remainder Theorem is true, divide f(x) by x - c to get

$$f(x) = (x - c)q(x) + r.$$

Then,

$$f(c) = (c - c)q(x) + r = 0 + r = r.$$

Thus the remainder is $f(c)$.

* * * *

Example 4.9

Find the remainder when $f(x) = 2x^3 - x^2 - 9$ is divided by $x - 3$.

Solution

$$f(3) = 2(3)^3 - 3^2 - 9 = 36.$$

Thus the remainder is 36.

* * * *

Example 4.10

Find the remainder when $g(x) = x^4 + 2x^3 + x^2 - 3x - 10$ is divided by $x + 2$.

Solution

$$g(-2) = (-2)^4 + 2(-2)^3 + (-2)^2 - 3(-2) - 10 = 0.$$

The remainder is 0.

* * * *

An important consequence of the Remainder Theorem is the Factor Theorem which gives a simple criterion for determining whether or not $x - c$ is a factor of a polynomial $f(x)$.

Factor Theorem   For any polynomial $f(x)$ and any real number c, $x - c$ is a factor of $f(x)$ if and only if $f(c) = 0$.

To see that the Factor Theorem is true, divide $f(x)$ by $x - c$,

$$f(x) = (x - c)q(x) + r.$$

Since the remainder theorem tells us that $r = f(c)$, we can write,

$$f(x) = (x - c)q(x) + f(c).$$

Clearly, $x - c$ is a factor of $f(x)$ if and only if the remainder $f(c)$ is zero.
* * * *

Example 4.11

Determine whether or not $x - 3$ is a factor of $2x^3 - x^2 - 9$.

Solution

Since $f(3) = 36$, $x - 3$ is not a factor of $f(x)$.

* * * *

Example 4.12

Determine whether or not $x + 2$ is a factor of $g(x) = x^4 + 2x^3 + x^2 - 3x - 10$.

Solution

Since $g(-2) = 0$, $x + 2$ is a factor of $g(x)$ according to the Factor Theorem.

* * * *

In summary,

given any polynomial f(x) and any real number c, x - c is a factor of f(x) if and only if c is a root of f(x).

We can now determine the maximum number of roots that a polynomial can have.   The product of n linear factors is a polynomial of degree n. Therefore, a polynomial of degree n can have at most n real roots.   If a polynomial has no linear factors then it has no roots.

Example 4.13

Determine the number of real roots for each of the following polynomial functions:

a. $f(x) = x^4 + 1$;

b. $g(x) = 5(x - 1)(x + 2)(x - 3)^2$.

Solution

a.   f has no linear factors and hence no real root.

b.   g has three distinct linear factors and thus three real roots.

* * * *

Exercises

In exercises 1 - 6, determine whether the given number c is a root of f(x).

1. $f(x) = x^3 + x^2 + x - 3, c = -1$

2. $f(x) = x^4 + 3x - 10, c = -2$

3. $f(x) = x^3 + 3x^2 - 4x - 12, c = 2$

4. $f(x) = x^4 - 2x^3 + x^2 - 36, c = 3$

5. $f(x) = 5 - 7x + 4x^3 - 3x^4, c = -1$

6. $f(x) = x(x-2)(x+3)+5, c = -3$

7.   Is x - 1 a factor of $f(x) = x^3 + x - 2$?

8.   Is x + 2 a factor of $f(x) = x^4 + 3x^3 + 5x + 18$?

9.   Write a polynomial which has degree 4 and real roots are 0, 1, and 3.

10.   Which of the following statements are true about the polynomial function

$$f(x) = ax^3 + bx^2 + cx + d?$$

Explain your answers.

a.   It has exactly three real roots.

b.   It has at least one real root.

c.   Its graph crosses the y-axis exactly once.

11.   True or false? If a polynomial P(x) has no real <u>roots</u> then either P(x) is always positive or P(x) is always negative. Explain your answer.

4.4   Electricity and Hot Water
(This section is optional, or can be done as an outside project.)

As the US population grows, so does its need for electricity. A coal burning power plant is put into operation on Chinle River (Section 4.1) at its entrance to the lake.   When coal is burned steam is produced, causing turbines to rotate and generate electricity. Only a portion (about 40%) of the heat produced is converted to electrical energy; the rest is waste heat which must be removed from the generator. Some of this heat is removed through the smokestack as hot gases (about 15%) and the remainder must be removed by other processes. In a common process, cooling occurs when water from a river or lake flows through the turbine condenser. The water is warmed by the waste heat and then is discharged back into the river. A large dependable flow of water must be available for this process, and the water used returns to the stream at a higher temperature, creating potentially harmful effects. In this section we will study the water flow necessary for this cooling process, how much the process increases the temperature of the stream from which it is drawn, and then determine what electrical generating capacity a plant on the Chinle River can support.

When heat is removed by water flowing through the condensers, the stream of water leaving has a higher temperature than the entering stream. Power (electrical energy) is measured in watts and work is measured in Joules. One watt equals one Joule per second. Therefore, if waste energy must be removed by cooling water at a rate of R megawatts, abbreviated MW, where 1 MW = 10^6 Joules per second then the energy discharged is

$$R \times 10^6 \text{ Joules.}$$

The specific heat of water (at 17 °C) is 4.184 Joules/gram degrees C; this means that the energy required to raise the temperature of 1 gram water 1° C is 4.184 Joules. Therefore it takes

$$4{,}184 \times M \times d \text{ Joules}$$

to raise the temperature of M grams of water d° C. Then R x 10^6 Joules will raise the temperature of

$$\frac{1}{4.184} \times \frac{R}{D} \times 10^6 \text{ grams} = 0.239 \times \frac{R}{d} \times 10^6 \text{ grams of water at d}° \text{ C.}$$

Therefore if the energy discharge is R megawatts and the temperature increase is to be d° C, then the required flow of water is $0.239 \times \frac{R}{d} \times 10^6$ grams per second. Since

$$10^6 \text{ grams} = 10^3 \text{ kilograms} = 1 \text{ cubic meter,}$$

this is $0.239 \times \frac{R}{d}$ cubic meters/sec.

If we convert this to cubic feet per second and degrees Fahrenheit, (using the conversion factors, 1 cubic meter = 35.31 cubic feet and 1° C = 1.8° F) then the relationship is

$$F = .239 \times 3531 \times 1.8 \frac{R}{T} = 152 \frac{R}{T} \ cfs,$$

or

$$F = 152 \frac{R}{T} \ cfs,$$

where

F = streamflow of the cooling water (in cubic feet per second),

R = rate at which waste heat is removed (in megawatts),

T = temperature change (in degrees Fahrenheit).

Example 4.14

A large coal burning power plant has the following characteristics:

*40% of the power generated is converted to electricity;

*15% of the waste heat is removed through the smokestack;

*85% of the waste heat is removed by cooling water;

*the temperature of the cooling water is increased 18ºF in the process.

If the generating capacity is 1000 megawatts, what streamflow is required for cooling? We use the following two pieces of information in solving this problem:

1000 MW = 10^6 kW = 10^9 x 3.412 Btu/hr =36x3.412 metric tons of coal per hour;

and

2500 MW = 307 metric tons per hour.

Solution

We begin by answering some easy questions.

1.   How much power does it take to produce 1000 MW electricity?

Let P be the total rate of power production in megawatts. The efficiency of the plant tells you what percent of the energy is converted to electricity. Since this plant has 40% efficiency and generates 1000 MW electricity, to answer that question, solve the equation .40P = 1000 to get P = 2500. Thus, the energy input (from burning coal) must be 2500 MW.

2.   At what rate is waste heat discharged to cooling water?

Since the total energy input is 2500 MW and 60% of that is converted to waste heat, then waste heat is produced at a rate of 1500 MW (60% of 2500). Eighty-five percent of the excess is removed by the cooling process, so the rate of discharge to cooling water is 85% of 1500 MW, which is 1275 MW.

3.   Determine the streamflow required to remove the waste heat, given that the temperature change is 18ºF.

We have R = 1275, T = 18; therefore

$$F = 1.52 \times \frac{1275}{18} = 1077 \text{ cfs(rounded)}$$

So, this generating plant requires a streamflow of 1077 cubic feet per second.

* * * *

This example shows that to generate 1000 MW electric power, at least 1077 cubic feet water per second must be diverted from the river. When the water is returned to the river, assuming no loss of water due to evaporation and no cooling, it will be 18°F warmer than the undiverted water in the river. If this warmer water contributes too substantially to the total flow, the river water temperature will increase to an unsatisfactory level. Excessive temperature changes will kill fish living in the water, increase evaporation, and have adverse long-term effects on the ecosystem. For example, the optimal temperature for brook trout is 58° F; the maximum average temperature desirable for sustaining growth is 65° F and a temperature of 78° F is lethal. For these and other reasons power plants are limited as to how much warm water can be returned to the river. In the next example we will see how much the water temperature downstream will increase from this warm water if the river flow is 3000 cfs.

Example 4.15

Assume sufficient water is withdrawn for cooling the generating plant described above and the warmed water is returned to the river. If the river has a flow of 3000 cfs, how much will the downstream river temperature increase? Assume the warm water is fully mixed with the undiverted water.

Solution

We have the same heat discharge, 1275 MW, but the flow is now considered to be the water flow of the entire river, 3000 cfs. Use the basic relationship

$$F = 1.52 \frac{R}{T} \text{ cfs}$$

and solve for T,

$$T = \frac{1.52 \times 1275}{3000} = 6.46$$

Therefore, the downstream water temperature will increase 6.46° F. Figure 4.3 illustrates the process.

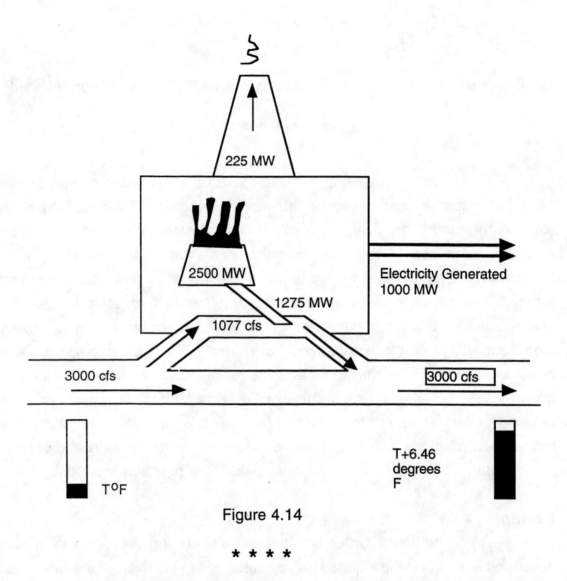

Figure 4.14

* * * *

Now we return to the Chinle River where the generating plant has been built ten miles downstream from the headwaters of the river. We will determine the maximum capacity for this plant and estimate the number of people that this electricity output can support.

Group Work

The coal-fired generating plant has the same characteristics as that described in the examples above:

*40% of the power generated is converted to electricity;

*15% of the waste heat removed through the smokestack;

*85% of the waste heat is removed by cooling water;

*the temperature of the cooling water is increased 18°F in the process. Use the streamflow of the river at the power plant determined in the previous section.

1. Determine the total rate of power production and the electric power generating capacity of the plant if all of the river water is used for cooling.

2. If all of the river water is used for cooling, the river temperature will increase 18° F, causing serious damage to the fish, plants and microorganisms in the river. To avoid this damage, a decision is made that only 30% of the flow can be diverted. With this restriction, determine the capacity of the plant and determine how much the river temperature will be increased after the warmed water is returned to the river.

3. It is estimated that a generating plant with capacity of 1000 MW will support 100,000 people and we will assume that the ratio of power to people is constant; i.e. that one megawatt supports 100 people. How many people can be supported by the generating plant with the restrictions described in part 2? How many without the restriction?

* * * *

CHAPTER FIVE
THE TOWN OF RIVER CITY

Introduction

We begin a study of water supply for a small but growing village, River City, on a mountain stream, Nizhoni River, which is fed primarily by snow. The village is located near the high elevation mountainous region which is the origin of the stream, and the entire water supply for the village comes from this stream. The situation is fictitious, but is designed as a sort of scaled-down version of the La Plata River which originates in the San Juan Mountains of southwestern Colorado, and a scaled-up version of the small town of Hesperus, Colorado, which is located very near the source of the La Plata. Due to the arid nature of this country it is very important to be able to predict the amount of water which will be available each year and when it will be available because during periods of high streamflow, it may be necessary to store water for use during drier periods. In this chapter, we study the population growth of the village and the resulting increase in water demand. In Chapter 8, we return to River City to study its future water needs; information obtained in this chapter will be used again there.

5.1 Population and Water Usage

The village of River City was started some years ago by a small group of people in search of solace. They were joined by others of like mind and presently the population is 3000 and has been growing at the rate of 6% per year. In order to predict water consumption, we must first create a model that predicts the village population in future years. Since the population is now 3000 with a projected growth rate of 6%, then next year there should be 3000 + .06(3000) people. Rather than calculate the numerical value of this expression we work to obtain a general function which will give the population in any year. So, factor out the common term 3000 to get

$$3000(1 + .06) = 3000(1.06)$$

people in the village 1 year from now. In 2 years there will be

$$3000(1.06) + .06[3000(1.06)] = 3000(1.06)[1 + .06] = 3000(1.06)^2$$

people.

Now follow this example in the first three problems below to obtain the desired function, then use your function to answer the remaining questions.

Group Work

1. Write the expression which gives the village population in 3 years; leave it as a product of terms as in the above example.

2. Repeat problem #1 for 4 years from now.

3. Generalizing from the examples and problems #1 and 2, write the function which gives the village population P(x) in x years from now (note that x = 0 in the current year). This function is called an *exponential function*; you will see more about exponential functions in the latter part of this chapter.

4. Graph the function P(x), $0 \leq x \leq 35$.

5. Predict the population of River City 5 years from now; 15 years from now; 30 years from now.

6. Use your calculator or computer to predict how many years before the population reaches 4000 people; 5000 people.

In the early days of River City, its residents' water needs were small; water for drinking, cooking, cleaning, etc., was all that was necessary. But as the town grew, people began trades, businesses were started, the need for community support developed, and some people began farming and raising livestock. With all of this, their per capita daily water consumption increased to about 195 gallons which is approximately 26 cubic feet. We will determine projected water needs based on this information and the population model.

7. Assuming a constant monthly amount of water is used each month, what is the total amount of water used monthly by the residents presently? Predict monthly usage in 30 years.

8. How many people will 50,000,000 cubic feet of water per month support? How many years before the population gets this large?

POPULATION SUMMARY

Fill in the required information and save it on a copy of this page; you will need this later. When writing in your answers, be sure to carefully explain the variables; this is important and you may forget them later on when you need them.

POPULATION FUNCTION P(x) (See question #4.)

5.2   Exponential Functions

The function derived in the preceding section to model the population is an exponential function. In the remaining sections of this chapter we will study these functions, their inverses, and their properties. The general form we use for an *exponential function* is

$$f(x) = a(b^x),$$

where a and b are constant, an x is the variable. The variable x is called the exponent, and the constant b is called the base. Restrictions on b are required,

b must be positive and different from 1.

(Note that if b = 1, then this function would be constant, and the restriction of b to positive values allows the domain of an exponential function to be all real numbers x.) Regardless of the value of x, $b^x > 0$, and hence the range of an exponential function will be

$$\{y| \, y > 0\} \text{ if } a > 0, \text{ or } \{y| \, y < 0\} \text{ if } a < 0.$$

Exponential functions are easily graphed on your calculator or computer; here are some examples.

Example 5.1

Graph $f(x) = 2^x$ (here, a = 1 and b = 2).

Solution

Using the calculator or computer, you should see the graph as shown in Figure 5.1. The restrictions on the axes for the portion of the graph shown are

$$-2 \le x \le 2; \, 0 \le y \le 10.$$

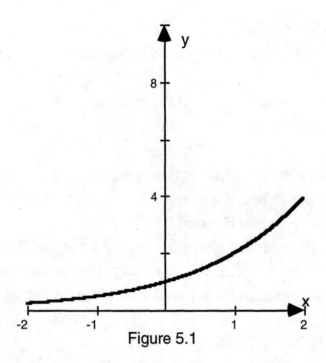

Figure 5.1

You should note that there are no x-intercepts; the y-intercept is (0, 1) which means $2^0 = 1$, and the entire graph is above the x-axis which means that $2^x > 0$ for all x.

* * * *

Example 5.2

Graph the functions $f(x) = 2^x$ and $g(x) = 10^x$ on the same coordinate system.

Solution

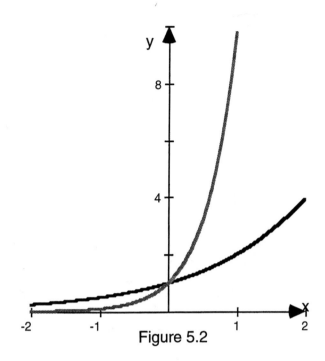

Figure 5.2

Refer to Figure 5.2. These graphs look basically the same except for the rate of increase. Note that the graph g(x) = 10ˣ is much steeper than the graph of f(x) = 2ˣ because $10 > 2$. In general, the larger the base, the steeper the graph, which means that the function increases really fast when the base is big.

* * * *

Example 5.3

    Graph $h(x) = (0.5)^x$.

Solution

    In this case $b = 0.5 < 1$ and the graph should appear as shown in Figure 5.3. This graph is decreasing, whereas the graph of $y = 2^x$ is increasing. If the base b is smaller than 1, the exponential function $y = b^x$ decreases, and if the base b is larger than 1, $y = b^x$ increases. Also, note from Figure 5.3 that $(0.5)^x > 0$. There are no x-intercepts and the y-intercept is (0, 1).

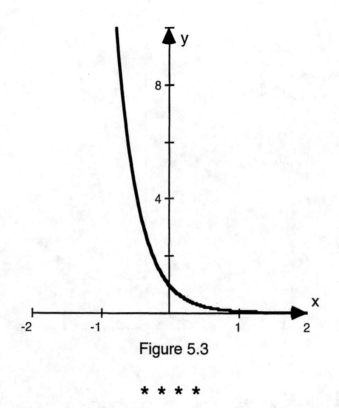

Figure 5.3

* * * *

Example 5.4

Graph both functions $h(x) = (0.5)^x$ and $y = .12^x$ on the same coordinate system.

Solution

You should see something like Figure 5.4. For these two functions, $y = .12^x$ decreases more rapidly.

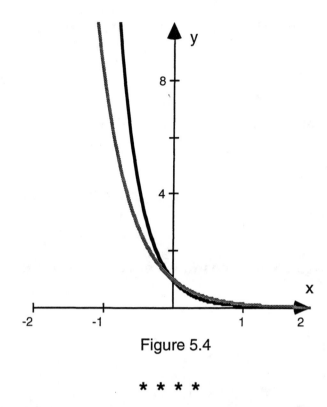

Figure 5.4

* * * *

In general, the graph of the exponential function $f(x) = b^x$ looks like one of the two shown in Figure 5.5.

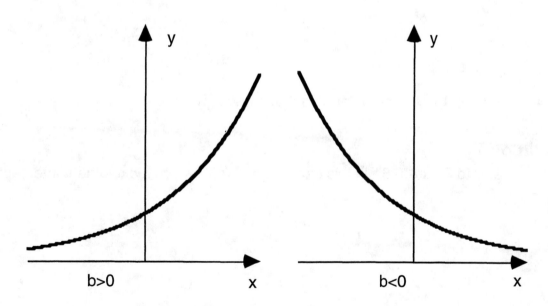

Figure 5.5

* * * *

We list some properties which are true for all exponential functions of the form $y = b^x$.

1.   There are no x-intercepts.

2.   The y-intercept is (0,1) since $b^0 = 1$.

3.   The entire graph is above the x-axis since $b^x > 0$ for all x.

4.   If $0 < b < 1$, the function decreases; the smaller the base b, the greater the rate of decrease.

5.   If $b > 1$, the function increases; the larger the base b, the greater the rate of increase.

Properties 6-10 are the ones which are commonly known as the *"laws of exponents."*

6. $b^p \times b^q = b^{p+q}$

7. $\dfrac{b^p}{b^q} = b^{p-q}$

8. $(b^p)^q = b^{pq}$

9. $b^{-p} = \dfrac{1}{b^p}$

10. $\dfrac{1}{b^{-p}} = b^p$

Example 5.5

Graph $y = 0.53(1.07)^x$ and determine the y-intercept.

Solution

Refer to Figure 5.6. In this example a = 0.53 which makes the y-intercept (0, 0.53).

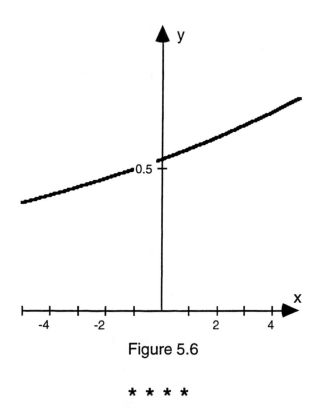

Figure 5.6

* * * *

Example 5.6

A very special number in mathematics is known as "e" with approximate value 2.71828. The number e occurs quite naturally in a variety of studies, in particular, in physics, biology, and even economics. The exponential function

$$f(x) = e^x$$

is known as the *natural exponential function*, its base is

$$b = e,$$

and its graph is shown in Figure 5.7. The notation "e" is for a famous mathematician, Leonard Euler (1707 - 1783).

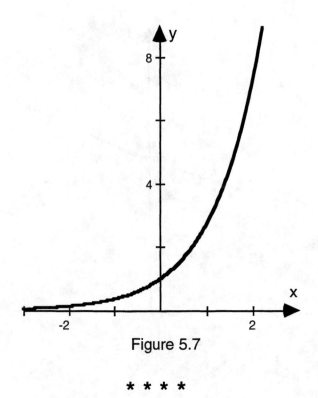

Figure 5.7

* * * *

Exercises

Evaluate each of the following.

1. $10^{.259}$
2. $(1.72)^{5.61}$
3. e^2
4. $3e^{2.1}$
5. $7.31(5.6)^{9.74}$

Graph each of the functions in Exercises 6-10 and determine the y-intercept.

6. $f(x) = 6^x$
7. $g(x) = -3^x$
8. $h(x) = (1.13)^x$
9. $k(x) = 9.6(4.3)^x$
10. $l(x) = -2.4e^x$

5.3   Logarithmic Functions

All exponential functions of the form $y = b^x, b > 0, b \neq 1$, have inverses because they are either strictly increasing (b > 1) or strictly decreasing (0 < b < 1) and hence are one-to-one. The inverse of the exponential function $f(x) = b^x$ is called the *logarithmic function with base b*. It is denoted by

$$f^{-1}(x) = \log_b x.$$

Since the domain of the exponential function consists of all real numbers, so will the range of the logarithmic function.   Similarly, the domain of the logarithmic function consists of all positive real numbers since this is the range of the exponential function.

By definition,

if $y = f(x) = b^x$ then $x = f^{-1}(y) = \log_b y$.

Therefore we see that

$$y = b^x \text{ if } x = \log_b y.$$

Similarly,

if $y = \log_b x$ then $b^y = x$.

Some examples should help your understanding.

Example 5.7

Determine $\log_2 8$.

Solution

The base in this case is 2 so

$$\log_2 8 = y \text{ if } 2^y = 8.$$

What power do you need to take 2 to in order to get 8? The answer is, of course, 3, since $2^3 = 8$. Therefore

$$\log_2 8 = 3.$$

* * * *

Example 5.8

Determine $\log_{10} 100$.

Solution

Here $\log_{10} 100 = y$ if $10^y = 100$. The answer is 2 because $10^2 = 100$, so

$$\log_{10} 100 = 2.$$

In this example, if we use the functional notation $f(x) = \log_{10} x$, then

$f(100) = 2.$

* * * *

Example 5.9

Determine $\log_{10} 0.1$.

Solution

You have to do a little more work here. You must find the exponent in the equation

$10^y = 0.1$.

The number 0.1 is the same as $\dfrac{1}{10}$, which is the same as 10^{-1}, so you have the equation

$10^y = 10^{-1}$

so $y = -1$; that is, $\log_{10} 0.1 = -1$.

* * * *

We now list the *properties of logarithmic functions* of the form $f(x) = \log_b x$. These will be needed later when we solve both exponential and logarithmic equations.

1. There are no y-intercepts regardless of the base b.
2. The x-intercept is $(1, 0)$; $\log_b 1 = 0$ since $b^0 = 1$.
3. The entire graph is to the right of the y-axis because the domain consists of only positive numbers.
4. $\log_b (MN) = \log_b M + \log_b N$.
5. $\log_b \dfrac{M}{N} = \log_b M - \log_b N$,
6. $\log_b (M^p) = p \log_b M$.
7. $\log_b b = 1$.
8. $\log_b 1 = 0$.
9. $\log_b b^x = x$.
10. $b^{\log_b x} = x$.

These properties can be derived from corresponding properties of exponents.

We will now examine the graphs of logarithmic functions with base 10 and base e.

Example 5.10

Graph $f(x) = \log_{10} x$.

Solution

The function $\log_{10}x$ is called the *common logarithmic function,* and is the inverse of the function 10^x; it is usually denoted by

$f(x) = \log x.$

Use the calculator or computer to graph this function. Figure 5.8 shows the portion of the graph with restricted intervals $0 < x < 10$ and $-2 < y < 2$.

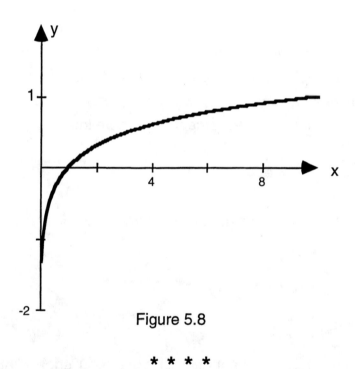

Figure 5.8

* * * *

The function $\log_e x$ is called the *natural logarithmic function.* It is the inverse of the function e^x and is written

$f(x) = \ln x.$

The graph of $y = \ln x$, which looks very much like $y = \log x$, is shown in Figure 5.9.

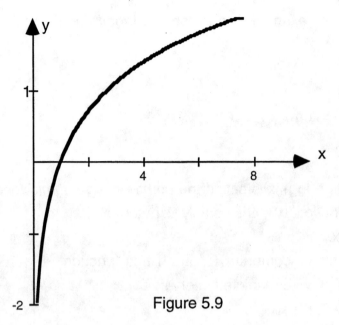

Figure 5.9

The domain of y = ln x consists of all x > 0, and the range consists of all real numbers y. The graph doesn't go to the left of the y-axis; it increases from −∞ to ∞, but it increases very, very slowly. Its x-intercept is (1, 0) because e^0 = 1 and hence ln = 0; it has no y-intercept. Also note that

ln e = 1

since ln e = $\log_e e$ and e^1 = e.

We now look at some graphs of logarithmic functions.

Example 5.11
Graph y = 3 + 2 ln x.

Solution
Use the calculator or computer. Figure 5.10 shows this graph. We estimate the x-intercept to be (.22, 0). There is no y-intercept.

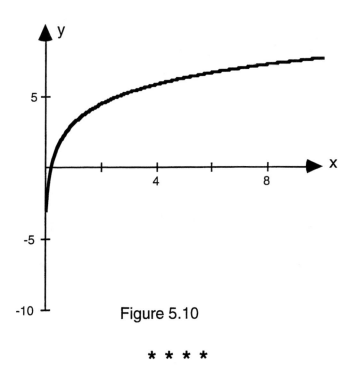

Figure 5.10

**   *   *   ***

Example   5.12

Graph y = 2.2 ln x - 1.7.

Solution

Figure 5.11 shows this graph. The x-intercept is (2.17, 0) rounded, and again, there is no y-intercept.

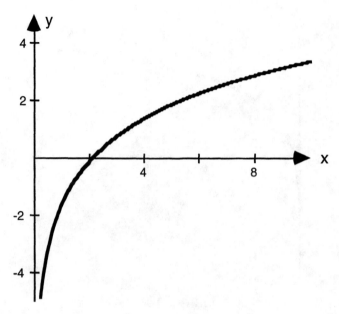

Figure 5.11

Example 5.13

Graph y = 2.3 - 4.5 ln x.

Solution

Figure 5.12 shows this graph. The x-intercept is (1.67, 0) rounded, and again, there is no y-intercept.

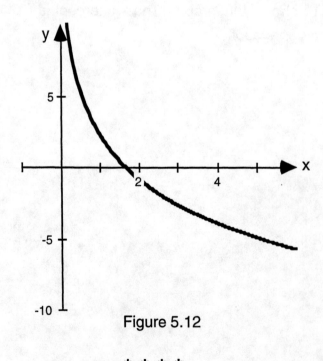

Figure 5.12

* * * *

Exercises

In exercises 1-7 determine the logarithms; use the calculator or computer for exercises 4-7.

1. $\log_3 81$
2. $\log_{10} .0001$
3. $\log_{10} 10000$
4. $\log 17$
5. $\log .023$
6. $\ln 2$
7. $\ln .03$

In exercises 8-12, graph the given function; use the calculator or computer to estimate the x-intercept for each.

8. $f(x) = 3 + 5 \log x$
9. $f(x) = 5 - 2 \log x$
10. $f(x) = 2 \ln x$
11. $f(x) = 1 - 3.2 \ln x$
12. $f(x) = 5 + 2.3 \ln x$

★ ★ ★ ★

5.4 Logarithmic and Exponential Equations

Consider Example 5.11 in Section 5.3. Using the calculator or computer we estimate the x-intercept of this graph to be x = 1.97. To find this intercept algebraically, it is necessary to solve the equation

$$0 = 2.2 \ln x - 1.7.$$

First solve for $\ln x$,

$$1.7 = 2.2 \ln x,$$
$$\ln x = 0.7727.$$

Therefore $e^{\ln x} = e^{0.7727}$, and recalling from Property 10 of logarithms that $e^{\ln x} = x$, we have

$$x = e^{0.7727} = 2.1656, \text{ rounded to four places.}$$

Example 5.14

Let $f(x) = 1.6 \ln x + 4$, and algebraically determine the x-intercept.

Solution

Solve the equation

$0 = 1.6 \ln x + 4$

to get

$-4 = 1.6 \ln x,$

or

$\ln x = -2.5.$

Therefore

$e^{\ln x} = e^{-2.5}$

and

$x = e^{-2.5} = 0.0821.$

* * * *

Example 5.15

Let $f(x) = 6 + 8.1 \ln x$, Determine x if $f(x) = 4$.

Solution A

Use the calculator or computer. Estimate the x-value when $f(x) = 4$. To do this, graph $y = 6 + 8.1 \ln x$ and the horizontal line $y = 4$ on the same coordinate system. (See Figure 5.13.)

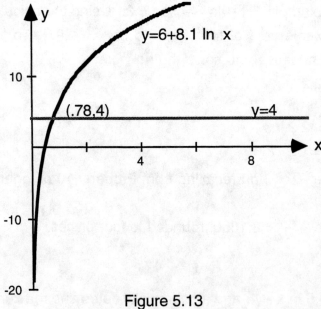

Figure 5.13

The x-value we want to estimate is the x-coordinate of the point of intersection of these two graphs. We estimate this to be x = .78.

Solution B
Solve for x algebraically when f(x) = 4. We solve the equation

$$4 = 6 + 8.1 \ln x$$
$$-2 = 8.1 \ln x$$
$$\ln x = -0.2469,$$

so

$$x = e^{-0.2469} = 0.7812.$$

Note that the point of intersection of the graphs is (0.7812, 4).

* * * *

Example 5.16
Solve the exponential equation
$$e^{2x+3} = 7.$$

Solution A
Use the calculator or computer. Let $f(x) = e^{2x+3}$ and estimate the x-value when f(x) = 7. Graph $y = e^{2x+3}$ and the horizontal line y = 7 on the same coordinate system. (See Figure 5.14.) The x value we want to estimate is the x-coordinate of the point of intersection of these graphs. We estimate this to be
$$x = -0.53.$$

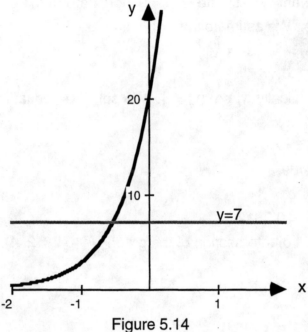

Figure 5.14

Solution B

Solve for x algebraically when $f(x) = 7$.

$$7 = e^{2x+3}$$

Take the natural logarithm of each side to get

$$\ln 7 = \ln (e^{2x+3}).$$

Recall from Property 9 that $\ln (e^z) = z$, so we get

$$\ln 7 = 2x + 3.$$

Next, evaluate $\ln 7$ on the calculator or computer and solve for x,

$$1.9459 = 2x + 3$$
$$2x = -1.0541$$
$$x = -0.5270.$$

* * * *

Example 5.17

Solve the exponential equation

$$2(1.2^x) = 5.$$

Solution

First divide by 2 to isolate the "exponent part."

$(1.2)^x = 2.5$;

Next take the natural logarithm of each side:

$\ln (1.2)^x = \ln (2.5)$;

then recall Property 9 of logarithms which says that in $b^x = x \ln b$. This gives

$x \ln (1.2)^x = \ln (2.5)$;

Now, we can solve for x,

$$x = \frac{\ln(2.5)}{\ln(1.2)} = 5.0257.$$

＊ ＊ ＊ ＊

Exercises

In each of the following,

A. graph each function on your calculator and use the graph to solve the given equation, and

B. solve each equation algebraically.

1. $f(t) = 1.2 + 3.7 \ln t$ solve $1.2 + 3.7 \ln t = 9$
2. $y = 500 + 225 \ln x$ solve $500 + 225 \ln x = 100$
3. $H(s) = .002 - .035 \ln s$ solve $.002 - .035 \ln s = .04$
4. $g(x) = e^{x+1}$ solve $e^{x+1} = 2$
5. $r(t) = e^{8t-4}$ solve $e^{8t-4} = 1$
6. $p(t) = 1.7(3^t)$ solve $1.7(3^t) = 4.5$

＊ ＊ ＊ ＊

5.5 Inverses of Exponential and Logarithmic Functions

The technique for finding the inverse of an exponential or logarithmic function is the same as that for a linear function, but a little more difficult.

Example 5.18

Determine the inverse of $f(x) = 2e^x$.

Solution

Replace f(x) by y to get

$y = 2e^x$.

Note that you need to solve for x, which is the exponent, and Property 9 of logarithmic functions enables you to do this. First divide by 2 to get

$$\frac{y}{2} = e^x,$$

then take the logarithm of each side to get

$$\ln\frac{y}{2} = \ln(e^x) = x\ln e.$$

Recall from Section 7.4 that ln e = 1 so that

$$\ln\frac{y}{2} = x.$$

Thus,

$$f^{-1}(y) = \ln\frac{y}{2}.$$

$$\ast\ \ast\ \ast\ \ast$$

Example 5.19

Determine the inverse of f(x) = e^{2x+3}.

Solution

Here,

y = e^{2x+3}

ln y = ln e^{2x+3}

ln y = 2x + 3.

Solve for x,

$$\ln y - 3 = 2x$$

$$x = \frac{\ln y - 3}{2}.$$

So

$$f^{-1}(y) = \frac{\ln y - 3}{2}.$$

To illustrate the relationship between f and its inverse, we observe

$$f^{-1}(f(x)) = f^{-1}(e^{2x+3}) = \frac{\ln(e^{2x=3}) - 3}{2}$$

$$= \frac{(2x+3)\ln e - 3}{2}$$

$$= \frac{2x+3-3}{2}$$

$$= \frac{2x}{2} = x.$$

Without computation, do you know what $f^{-1}(f(2))$ is?

✱ ✱ ✱ ✱

Example 5.20
Determine the inverse of $f(x) = 2^x$.

Solution
While the definition of a logarithmic function says that this is $\log_2 y$, there is no $\log_2$ key on most calculators, so for convenience, we illustrate how the inverse of $f(x) = 2^x$ can be expressed in terms of natural logarithms. Set
$$y = 2^x$$
and solve for x by taking the natural logarithm of both sides. We get
$$\ln y = \ln 2^x$$
$$\ln y = x \ln 2 = x(.693),$$
and
$$x = \frac{\ln y}{.693}.$$
Therefore,
$$f^{-1}(y) = \frac{\ln y}{.693}.$$

✱ ✱ ✱ ✱

Example 5.21
Determine the inverse of $f(x) = 2.35(1.1^x)$.

Solution
Set $y = 2.35(1.1^x)$ and solve for x. First you must divide by 2.35 to get
$$\frac{y}{2,35} = 1.1^x;$$
then take logarithms and get
$$\ln(\frac{y}{2.35}) = \ln(1.1^x)$$
$$= x \ln 1.1$$
$$= x(0,95).$$
This gives

$$x = \frac{1}{.095}\ln(\frac{y}{2.35}).$$

* * * *

The following examples illustrate techniques for finding inverses of natural logarithmic functions. You will need to recall these properties of logarithms,

$$e^{\ln x} = x \text{ and } \ln e^x = x.$$

Example 5.22

Determine the inverse of f(x) = 2 + 3 ln x.

Solution

Replace f(x) by y to get

y = 2 + c ln x.

Now solve for ln x,

$$\ln x = \frac{y-2}{3};$$

then solve for x,

$$e^{\ln x} = e^{\frac{y-2}{3}},$$

$$x = e^{\frac{y-2}{3}},$$

and finally,

$$f^{-1}(y) = e^{\frac{y-2}{3}}.$$

We check our work by forming both composites ,

$$f(f^{-1}(y)) = 2 + 3\ln(e^{\frac{y-2}{3}})$$

$$= 2 + 3\frac{(y-2)}{3}$$

$$= y,$$

and

$$f^{-1}(f(x)) = e^{\frac{(2+3\ln x)-2}{3}} = e^{\ln x} = x,$$

just as it should be.

* * * *

Example 5.23

Determine the inverse of $g(x) = 3.7 + 5 \ln (2x)$.

Solution

Set

$$y = 3.7 + 5 \ln (2x),$$

and solve for $\ln (2x)$,

$$\ln 2x = \frac{y - 3.7}{5}.$$

Therefore

$$e^{\ln(2x)} = e^{\frac{y-3.7}{5}},$$

$$2x = e^{\frac{y-3.7}{5}},$$

$$x = \frac{1}{2} e^{\frac{y-3.7}{5}}.$$

Finally,

$$g^{-1}(y) = \frac{1}{2} e^{\frac{y-3.7}{5}}.$$

Now form the composites to check our work,

$$g(g^{-1}(y)) = 3.7 + 5 \ln(2(\frac{1}{2})e^{\frac{y-3.7}{5}})$$

$$= 3.7 + 5 \ln e^{\frac{y-3.7}{5}}$$

$$= 3.7 + 5(\frac{y-3.7}{5}) = y;$$

and

$$g^{-1}(g(x)) = \frac{1}{2} e^{\frac{3.7+5\ln(2x)-3.7}{5}}$$

$$= \frac{1}{2} e^{\ln(2x)}$$

$$= \frac{1}{2}(2x)$$

$$= x.$$

* * * *

Exercises

In exercises 1-3, determine the inverse of the given exponential function by using natural logarithms; for exercises 4-5, determine the inverse of the given logarithmic function. Then form both composites to show that $f^{-1}[f(x)] = x$ and $f[f^{-1}(y)] = y$, and compute the indicated functional values.

1. $g(x) = 3^x$. Compute $g(2)$ and $g^{-1}(9)$.
2. $h(x) = 2.1\,(7.01)^x$. Compute $h(4.3)$ and $h^{-1}(9095.085205)$.
3. $f(x) = (6.03)^x$. Compute $f(2.5)$ and $f^{-1}(2.5)$.
4. $f(x) = 2.7 + 1.3 \ln x$; compute $f(2)$ and $f^{-1}(3.601)$.
5. $g(x) = 5 - 3 \ln x$: compute $g(7)$ and $g^{-1}(-.8377)$.

* * * *

CHAPTER SIX
TRIGONOMETRIC FUNCTIONS

6.1   Angles and Triangles

There are six trigonometric functions which will be formally introduced in this chapter as functions of real numbers. They are called sine, cosine, tangent, secant, cosecant, and cotangent. Before we learn their formal definitions, we will show how some of these functions can be used to study right triangles. First, we should know some general properties of all triangles; for this, refer to the triangle shown in Figure 6.1. We use a rather common practice for labeling triangles. The vertices are labeled with capital letters, the side opposite each vertex is labeled with the corresponding lower case letter, and the angle itself with the corresponding Greek letter. However, sometimes the Greek letters are suppressed and the capital letter for the vertex is also used to denote the angle.

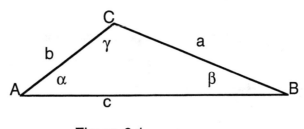

Figure 6.1

One common unit of measure of an angle is the *degree*. If a circle is divided into 360 equal parts by radii emanating from its center, then the measure of the angle formed by any two adjacent radii is *one degree*. An angle of measure 90° is a *right* angle, an angle of measure less than 90° is an *acute angle*, and an angle with measure greater than 90° is called a *obtuse angle.* In Figure 5.1, the angles α and β are acute whereas γ is obtuse. The sum of all the interior angles in any triangle is always 180°, regardless of its shape or size, that is,

$\alpha + \beta + \gamma = 180°.$

A triangle which has one of its interior angles equal to 90° is called a *right triangle*; see Figure 6.2. Here $\gamma = 90°$ and therefore we see that in any right triangle, the sum of the other two angles will always be 90°, i. e., in the right triangle shown, $\alpha + \beta = 90°$. The side opposite the right angle is called the *hypotenuse*, note that the hypotenuse must always be longer than either of the other two sides.

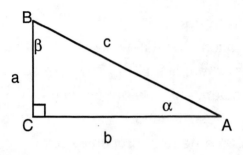

Figure 6.2

The sides of any right triangle enjoy a very special relationship, this is given in the famous Pythagorean Theorem:

A triangle is a right triangle if and only if the sum of the squares of the lengths of two of its sides equals the square of the length of the third side.

When we have a right triangle, the specific relationship among the sides is this: *the square of the length of the hypotenuse is equal to the sum of the squares of the lengths of the other two sides.* In Figure 6.2,

$a^2 + b^2 = c^2$.

The Pythagorean theorem works two ways for us. If we know that we have a right triangle, then we have the stated relationship among the sides; on the other hand, we can decide whether or not a particular triangle is a right triangle by checking if the relationship holds. The following examples illustrate how these work.

Example 6.1

In right triangle ABC shown in Figure 6.3, a= 5.2 and b = 12.9. Find the length of the hypotenuse c.

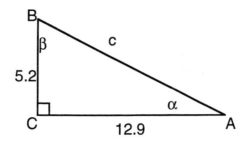

Figure 6.3

Solution

We use the Pythagorean Theorem with a = 5.2 and b = 12.9,

$$5.2^2 + 12.9^2 = c^2.$$

Simplifying the left hand side gives

$$27.04 + 166.41 = c^2$$

$$193.45 = c^2.$$

Taking the square root of both sides we get c = 13.9 (to the nearest tenth).

* * * *

Example 6.2

In right triangle ABC shown in Figure 6.4 below, a = 4.9 and c = 14.2. Find b.

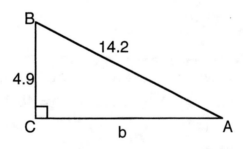

Figure 6.4

Solution

We use the Pythagorean Theorem with a = 4.9 and c = 14.2,

$$4.9^2 + b^2 = 14.2^2$$

Simplify and solve for b,

$$24.01 + b^2 = 201.64$$
$$b^2 = 177.63$$
$$b = \sqrt{177.63} = 13.3 \text{ (to the nearest tenth).}$$

* * * *

Example 6.3

Determine whether or not each of the triangles with the given sides is a right triangle.

a. $a = 1.00, b = 2.40, c = 2.60$;

b. $a = 3, b = 6, c = 8$.

Solution

a. $a^2 = 1.0^2 = 1.0;\ b^2 = 2.4^2 = 5.76;\ c^2 = 2.6^2 = 6.76$.

Since $1.0 + 5.76 = 6.76$ the triangle is a right triangle by the Pythagorean Theorem.

b. $a^2 = 3^2 = 9; b^2 = 6^2 = 36; c^2 = 8^2 = 64$.

Since no two squares add up to the third, the triangle is not a right triangle.

* * * *

Class Work

1. In right triangle ABC shown in Figure 6.5, a = 70.5 and b = 42.1. Find the length of the hypotenuse c.

2. In right triangle ABC shown in Figure 6.5, b = 10.72 and c = 14.45. Find a.

3. Is the triangle with sides a = 9.0, b = 12.0 and c = 45.0 a right triangle?

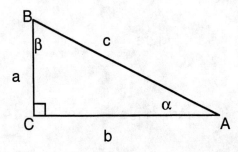

Figure 6.5

* * * *

Exercises

In each of the right triangles shown below, find the missing side. Give your answers to the nearest tenth.

1. $a = 26.3$, and $b = 16.4$, find c.

2. $b = 3.4$, and $c = 5.7$, find a.

3. $a = 2998$, and $c = 3256$, find b.

4. $a = 0.6$, and $b = 0.3$, find c.

Determine whether the triangle with the given sides is a right triangle.

5. $a = 3.2$, $b = 7.1$, and $c = 8.5$.

6. $a = 5$, $b = 5$, and $c = 5\sqrt{2}$.

7. $a = 21$, $b = 24$, and $c = 28$.

$$* \quad * \quad * \quad *$$

6.2   Right Triangle Trigonometry

Refer to the Figure 6.6 shown below. The functions sine, cosine, and tangent of the angle α (abbreviated $sin\alpha$, $cos\alpha$, and $tan\alpha$) are defined by:

$$sin\alpha = \frac{a}{c};$$

$$cos\alpha = \frac{b}{c};$$

$$tan\alpha = \frac{a}{b}.$$

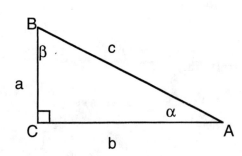

Figure 6.6

These functions of either acute angle in any right triangle can described as follows:

$$\text{sine of angle} = \frac{\text{side opposite}}{\text{hypotenuse}};$$

$$\text{cosine of angle} = \frac{\text{side adjacent}}{\text{hypotenuse}};$$

$$\text{tangent of angle} = \frac{\text{side opposite}}{\text{side adjacent}}.$$

In Figure 6.6 above:

$$sin\beta = \frac{b}{c};$$

$$cos\beta = \frac{a}{c};$$

$$tan\beta = \frac{b}{a}.$$

The following examples illustrate the use of these functions in right triangle problems. (Make sure your calculator mode is set on degrees.)

Example 6.4

If $\alpha = 37°$ and b = 14.3, determine a.

Solution

We use the tangent function since the sides involved are opposite and adjacent to the known angle.

$$\tan 37 = \frac{a}{14.3}$$

$$a = 14.3 \cdot \tan 37 = 10.78.$$

* * * *

Example 6.5

If $\beta = 22.6°$ and c = 41.73 determine the remaining sides and angles.

Solution

We use the sine function to determine b since the sides involved are the hypotenuse and the one opposite the known angle,

$$\sin 22.6 = \frac{b}{41.73}$$

$$b = 41.73 \cdot \sin 22.6 = 16.04.$$

Although it is possible to use the Pythagorean Theorem to determine a, we use the cosine to further illustrate the use of the trigonometric functions,

$$\cos 22.6 = \frac{a}{41.73}$$

$$a = 41.73 \cdot \cos 22.6 = 38.53.$$

Of course, $\gamma = 90^\circ$ and since the sum of the measures of the angles of a triangle is 180 degrees then $\alpha = 67.4^\circ$.

∗ ∗ ∗ ∗

Example 6.6

If a = 98.6 and b = 212, determine α.

Solution

We use the tangent function since the known sides are opposite and adjacent to angle α, $\tan \alpha = \frac{98.6}{212}$. Use the inverse tangent key (tan^{-1}) on your calculator to determine the angle whose tangent equals $\frac{98.6}{212}$,

$$\alpha = \tan^{-1}\left(\frac{98.6}{212}\right) = 24.94^\circ.$$

∗ ∗ ∗ ∗

Example 6.7

If b = 6.89 and c = 8.4 determine β.

Solution

In this case we use the sine function since the known sides are the hypotenuse and the one opposite to angle β,

$$\sin \beta = \frac{6.89}{8.4}$$

$$\beta = \sin^{-1}\left(\frac{6.89}{8.4}\right) = 55.12^\circ$$

∗ ∗ ∗ ∗

147

Class Work

Refer to the right triangle ABC shown above in Figure 6.6. The angles α, β and γ are measured in degrees and γ = 90°.

1.   If a = 18 and c = 20.5, find α.
2.   If a = 11.6 and b = 8.3, find ß.
3.   If b = 112.4 and c = 234.5, find a.

* * * *

Exercises

In exercises 1-5, refer to the right triangle ABC shown below. The angles α, β, and γ are measured in degrees and γ = 90°.

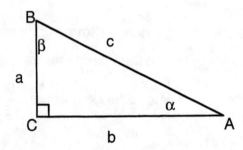

1.   If β = 23.5° and a = 15.7, determine b.
2.   If α = 33.2° and c = 110.76, determine the remaining sides and angles.
3.   If a = 18.6 and b = 27.2, determine α.
4.   If a = 90.1 and c = 131.7, determine α.
5.   If b = 16.89 and c = 31.56, determine α.
6.   A right triangle contains a 32° angle. If one leg is 4.5 inches, what is the length of the hypotenuse. (Two answers are possible.)
7.   A right triangle has a hypotenuse of length 6.2 cm. If one angle is 43°, find the length of each of the other sides.
8.   Find the height of the Great Pyramid of Egypt, using the information given in the figure below.

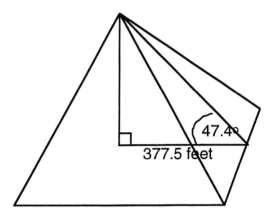

9.   Find the distance from point P to point Q across the river shown in the figure below.

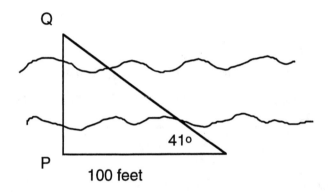

10.   At 13,680 ft. from the top of Mauna Loa, an active volcano in the Hawaiian Islands, the angle of depression of the horizon on the Pacific Ocean is 2.07°. Use this information to approximate the radius of the earth. For greater accuracy, determine the radius in feet rounding all intermediate answers to 5 decimal places, then convert your final answer to the nearest mile.

11.   If you are riding in a car in a rain storm, the raindrops appear to be falling at an angle instead of straight down. The faster you go, the greater the angle will be to the vertical. Generally if an object is moving toward or away from you in a direction perpendicular to your travel, then the apparent direction of travel of the object is dependent upon its speed and the speed you, as the observer, are moving. Refer to the figure below.

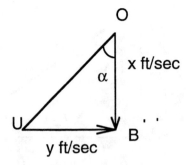

Suppose an object is moving vertically down at x ft./sec., and you are traveling at y ft./sec. Also assume the object is at 0 when you are at U, and that the object bonks you on the head at B. Then the angle α, measured from the vertical will be the apparent direction of travel of the object and can be determined using the tangent, $tan\alpha = \dfrac{x}{y}$, or $\alpha = tan^{-1}\dfrac{x}{y}$. The angle α is called the *aberration* of the object.

For example, if you are driving your car at 30 mph and a raindrop falling vertically at 10 mph hits your windshield, then the angle α, or the aberration of the raindrop, is determined by

$$\alpha = tan^{-1}\frac{30}{10} = 71.6^{\circ},$$

i.e., the apparent direction the raindrop is falling is 71.6º from the vertical. If you were holding a tube so that the raindrop would fall through it without touching the sides, you would have to hold it at an angle of 71.6º to the vertical.

This same phenomenon is true for light. Light travels at a constant speed and the earth is revolving about the sun at a constant speed, so light from a star appears to be displaced also. Thus, in order to view a star directly overhead through a telescope, the telescope must be tilted by the aberration of the star. The maximum aberration occurs when the path of the earth and the line of sight to the star are perpendicular. It has been observed that the maximum aberration for all stars is the same, $\alpha=20.47"$, and this is called the *constant of aberration*. This constant can be used to find the orbital velocity of the earth and the distance from the earth to the sun; follow the steps below:

A) The velocity of light is constant, 186273 mi./sec. Use this and the constant of aberration to determine the orbital velocity of the earth (to two decimal places) in mi./sec.

B) The earth completes one orbit of the sun every 325.25 days. Although the orbit is elliptical, you can assume a circular orbit to approximate the mean distance from the earth to the sun. Use the two formulas below to compute this distance.

> Distance = Rate x Time
> Circumference of a circle = π x Diameter

* * * *

6.3   Sine of the Times

Ocean tides are a cyclic phenomenon recurring approximately every twelve hours. Our goal is to determine a function which will predict the water level at any specific time of day. The following assumptions will apply to this beach:

* mid-tide occurs at 12 noon and at midnight;
* high tide is at 3 p.m.;
* low tide is at 9 p.m.;
* at high tide, the water level is 4 feet above the height at mid-tide;
* at low tide, the water level is 4 feet below the height at mid-tide.

This information can be summarized in Table 6.1 using t = 0 for noon and water level at y = 0 at mid-tide.

t	y
0	0
3	4
9	-4
12	0

Table 6.1

These data can be conveniently represented by a clock with one hand and radius 4 placed on an (x,y) coordinate system as shown in Figure 6.7.

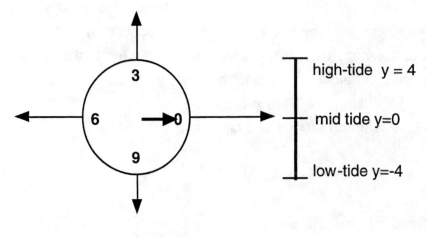

Figure 6.7

At time t = 0 the hand is on the positive x axis and rotates counterclockwise one complete revolution is 12 hours. The tip of the hand is always at the surface of the water. For example, in the next picture (Figure 6.8) it appears to be about 1:30 p.m. and the water level is approximately 2.8 feet above mid-tide.

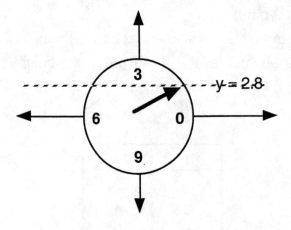

Figure 6.8

As the clock hand turns, it determines points on the coordinate system. The second coordinate y is the height of the water above or below the mid-tide mark at time t.   The tip of the hand traces the graph of a function which is a trigonometric function called the *sine function*. This sine function has amplitude 4 feet (maximum height above or below mid-tide); since the tide is cyclic, this pattern repeats every 12 hours, and we say that this sine function has period 12. (See Figure 6.9).

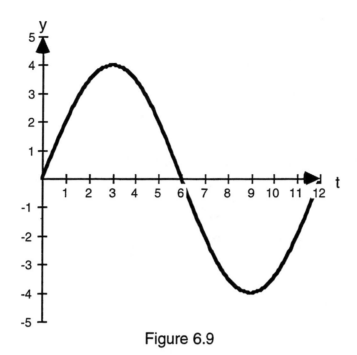

Figure 6.9

In order to indicate the water level at any time in the future or past we could place this basic graph end-to-end from the beginning to the end of time. (See Figure 6.10.)

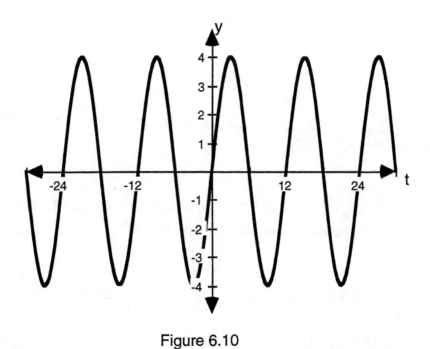

Figure 6.10

* * * *

6.4   Measurement of Angles

Before giving a formal definition of the sine function we need to provide more information about angles.   Other than the "degree", there is another very important unit of measure for angles, this is *radian* measure and must be used when defining the trigonometric functions of real numbers. A *central angle* in a circle is an angle with vertex at the center of the circle. A central angle in a circle which subtends an arc on the circumference of length equal to the radius has measure *one radian*. See Figure 6.11.

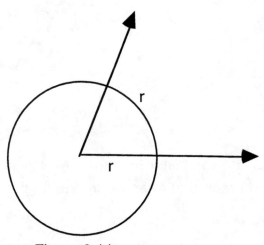

Figure 6.11

The ancient Greeks years ago discovered that the ratio of the circumference of a circle to its diameter is the same regardless of the size of the circle, this constant ratio is known as the number π. So if C is the circumference and d is the diameter of a circle, then

$$\pi = \frac{C}{d}$$

The number π is irrational and has an approximate value of 3.14159.

If r denotes the radius of the circle, then

$$\pi = \frac{C}{2r}.$$

From this, we get the formula for the circumference of a circle,

$$C = 2\pi r.$$

Now, if an angle of measure 1 radian subtends an arc of length r, and a 360° angle "subtends" an arc of length $2\pi r$, then the radian measure of 360° can be obtained by dividing the circumference by the radius, $\dfrac{2\pi r}{r} = 2\pi$. Hence we get the conversion formula,

$$360^0 = 2\pi \text{ radians.}$$

This can be simplified to

$$180^0 = \pi \text{ radians,}$$

and can easily be used to convert either way. To convert radians to degrees, divide both sides by π to get

$$1 \text{ radian } = \left(\dfrac{180}{\pi}\right)^0.$$

To convert degrees to radians, divide both sides by 180 and we have

$$1^0 = \left(\dfrac{\pi}{180}\right) \text{ radians.}$$

Example 6.8

An angle has measure 50^0. Find its radian measure.

Solution

Since $1^0 = \dfrac{\pi}{180} \, radians$, we can multiply by 50,

$$50^o = 50\left(\dfrac{\pi}{180}\right) \text{ radians} = \dfrac{5\pi}{18}\text{radians} = .87 \text{ radians (rounded to two}$$

decimal places).

* * * *

Example 6.9

An angle has measure $-\dfrac{\pi}{3}$ radians. Find its degree measure.

Solution

Since 1 radian $= \dfrac{180}{\pi}^{\circ}$, we can multiply by $-\dfrac{\pi}{3}$,

$$-\dfrac{\pi}{3} \text{ radians} = -\dfrac{\pi}{3}\left(\dfrac{180}{\pi}\right)^o = -\dfrac{180^0}{3} = \text{-60}^{\circ}.$$

* * * *

Example 6.10

An angle has measure 1.25 radians. Find its degree measure.

Solution

Since 1 radian = $\dfrac{180^{o}}{\pi}$, we can multiply by 1.25,

$1.25\left(\dfrac{180}{\pi}\right) = 71.62^{o}$ (rounded to two decimal places).

* * * *

Class Work

1. Convert each angle to radians. Express your answer in decimal form, rounded to two decimal places.

    a. 100°

    b. -36°

2. Convert each angle to degrees. Express your answer in decimal form, rounded to two decimal places.

    a. $\dfrac{\pi}{8}$ radians

    b. $-\dfrac{5\pi}{6}$ radians

    c. 2.5 radians

* * * *

6.5   Arc Length

In a manner similar to that used in deriving the conversion formula, we replace the 360° angle by an arbitrary central angle θ which subtends an arc of length S on the circle, the radian measure of θ will be

$$\frac{S}{r} = \theta,$$

and this gives a convenient formula for determining the length of an arc on a circle subtended by a central angle,

$$S = r\theta,$$

where it is important to remember that θ must be measured in radians. The following examples show how this can be used.

Example 6.11

Find the length of the arc of a circle of radius 4 inches, subtended by a central angle of 0.75 radians.

Solution

We use the equation $S = r\theta$ with $r = 4$ and $\theta = 0.75$,
$$S = 4(.75) = 3.0 \text{ inches.}$$

$$* \ * \ * \ *$$

Example 6.12

A central angle in a circle of radius 5 feet subtends an arc of length 3.6 feet. Find the measure of the angle.

Solution

Let θ represent the measure of the central angle and use the equation $S = r\theta$ with $S = 3.6$ and $r = 5$,
$$3.6 = 5\theta$$
The central angle is $\theta = 3.6/5 = .72$ radians.

$$* \ * \ * \ *$$

Example 6.13

A central angle of 110º in a circle subtends an arc of length 18.2 in. Find the radius of the circle.

Solution

Let θ denote the measure of the central angle. We must first express θ in radians,
$$\theta = 110\left(\frac{\pi}{180}\right) \text{ radians} = 1.92 \text{ radians.}$$
Next, we use the equation $S = r\theta$ with $S = 18.2$ and $\theta = 1.92$,
$$18.2 = r(1.92)$$
$$r = \frac{18.2}{1.92} = 9.5 \text{ (rounded to one decimal place).}$$

The radius is approximately 9.5 inches.

* * * *

Class Work

S denotes the length of an arc of a circle of radius r subtended by a central angle θ. Find the missing quantity.

1. r = 11.4 miles, θ = .5 radian, S = ?
2. θ = 105°, S = 6 inches, r = ?
3. r = 3.2 feet, S = 4 feet, θ = ?

* * * *

6.6   The Sine Function

Replace the clock in the example by a circle with center at the origin and radius r > 0. The radius should replace the hand of the clock and rotate from the positive x-axis. As the radius rotates from its initial position on the positive x-axis an *angle* is formed; the *initial side* is the positive x-axis, the *terminal side* is the radius, and the *vertex* is the origin. Counter-clockwise rotations generate positive angles, and by allowing the radius to also rotate in a clockwise direction, negative angles are generated. (See Figure 6.12.)

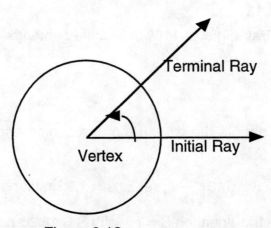

Figure 6.12

For example, after one-quarter counter-clockwise rotation, the terminal side is the positive y-axis and the measure of the resulting angle is 90°; one

158

complete counter-clockwise rotation corresponds to an angle of 360º, and two complete clockwise rotations gives an angle with measure -720º. If the radius is not rotated from the initial position then the angle has measure 0º. (See Figure 6.13.)

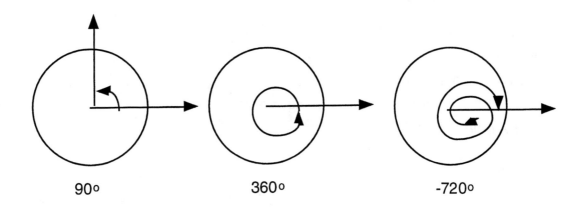

90º   360º   -720º

Figure 6.13

In order to define the sine function, we return to the circle with center at the origin and radius r > 0. (See Figure 6.14.) Let t be the radian measure of an angle and let (x, y) be the point of intersection of the terminal side of the angle with the circle. The sine function, abbreviated sin, is defined by

$$\sin t = \frac{y}{r}.$$

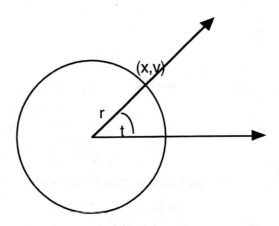

Figure 6.14

This definition is independent of the size of the circle. This can be easily seen by consideration of similar triangles. Refer to Figure 6.15. Suppose we use two circles, one of radius r and another of radius r', to define the sine of a real number t, $0 < t < \frac{\pi}{2}$. Then we get

$$\sin t = \frac{y}{r} \text{ and } \sin t = \frac{y'}{r'}.$$

Of course, if the sine is indeed a function, these should be equal. The two triangles on the right are formed inside the two circles. It should be clear that these two triangles are similar: they have one angle in common, one right angle each, and therefore corresponding angles are equal. Hence ratios of corresponding sides are equal, in particular,

$$\frac{y}{y'} = \frac{r}{r'}, \text{ or } \frac{y}{r} = \frac{y'}{r'},$$

i. e., the "two" values for sin t are the same. A similar argument can be used to prove this for any angle t.

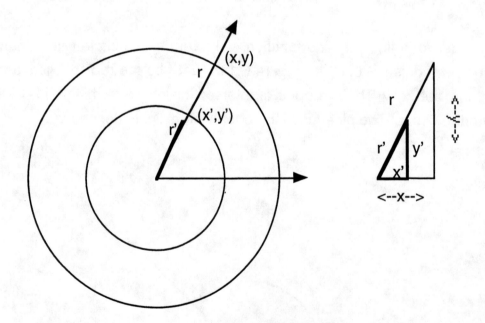

Figure 6.15

Note that the sine is a function of t and t is a real number, the radian measure of an angle. Also, given any real number t, there is an angle with measure t radians, therefore the domain of the sine consists of all real numbers. When the circle has radius 1 (the unit circle) then we have $\sin t = y$.

In order to see the graph of the sine function, it is convenient to visualize the radius rotating counterclockwise once around the unit circle from its initial position as indicated by Table 6.2 below.

As t varies from	sin t varies from
0 to π/2	0 to 1
π/2 to π	1 to 0
π to 3 π/2	0 to -1
3π/2 to 2π	-1 to 0

Table 6.2

On a coordinate system with horizontal axis t and vertical axis y, the graph of y = sin t over the interval [0, 2π] is shown. See Figure 6.16.

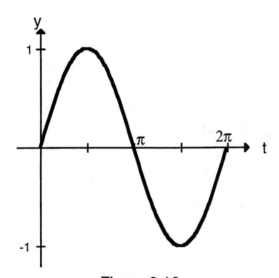

Figure 6.16

If the terminal side continues to rotate in either a clockwise or counter clockwise direction around the circle, it should be clear that the values of the sine simply repeat over its entire domain (the real line). A portion of the graph of y = sin t is as shown in Figure 6.17.

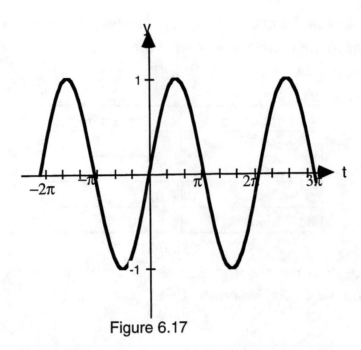

Figure 6.17

Note that the maximum value is 1, the minimum value is -1, so the range of the sine function is the closed interval [-1, 1]. The sine function is periodic, i.e., its values repeat over certain adjacent intervals of equal length. The length of the smallest such interval is the period, and in this case is 2π.

Class Work

Use your calculator or computer to graph each of the following functions. Show at least two complete cycles.

1. y = 2 sin t
2. y = sin 3 t
3. y = sin (t + 4)
4. y = 5 + sin t
5. y = 5 + 2 sin (3(t + 4))

How would you change the graph of y = sin t to get each of the above graphs?

* * * *

6.7   Features Of The Graphs Of The Sine Function

We consider four important features of the graph of a sine function of the form y = a + bsin(ct + d).

1)   *Vertical Shift.* The graph of the function y = a + bsin t is obtained by shifting the graph of y = sin t vertically a units from the t-axis. Thus, the *vertical shift* is a. Compare the graphs of y = sint, and y = 2 + sint in Figure 6.18 a. Note that the first graph has a horizontal "centerline" on the x-axis, i. e., the graph goes the same distance above the t-axis as it goes below. However the "centerline " of the second graph is the line y = 2. The vertical shift of y = sint is 0 whereas the vertical shift of y = 2 + sint is 2 units above the t-axis. The vertical shift can be either positive or negative; the graph of y = -2 + sint has vertical shift -2 and is shifted 2 units down (see Figure 6.18 b).

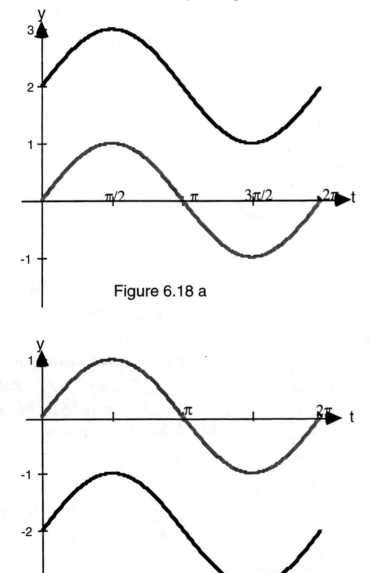

Figure 6.18 a

Figure 6.18 b

2)   *Amplitude*. The *amplitude* is the absolute value of the coefficient of sine, i. e., |b|. Compare the graphs of y = sint and y = 3sint in Figure 6.19 below. Note that the first graph has amplitude 1 and goes 1 unit above and 1 unit below the t-axis whereas the second graph goes 3 units above and below its centerline, the t-axis. The first function has range [-1, 1] and the second has range [-3, 3]. Note that the amplitude is always positive.

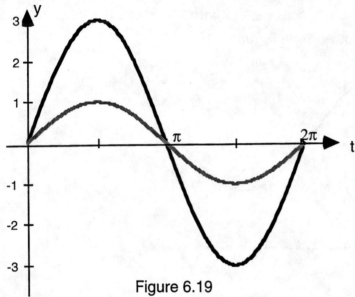

Figure 6.19

3)   *Horizontal Shift*. If the graph of a trigonometric function is shifted horizontally h units, then the number h is called the *phase shift*. Compare the graphs of y = sint and the other sine graph shown in Figure 6.20 below. The second graph is also one cycle of the sine, but it "starts" at t = 1 instead of t = 0; this sine function has a phase shift of +1. The phase shift can be either positive or negative; a positive phase shift indicates a shift to the right, a negative indicates a shift to the left. We will show you how to find the phase shift in examples that follow.

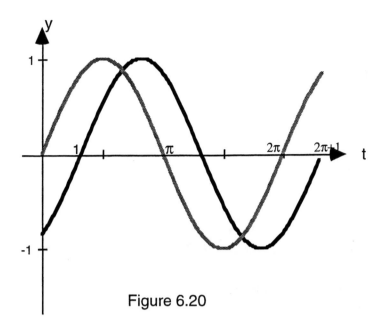

Figure 6.20

4)   *Period.* A function is *periodic* if there is a number p > 0 such that
        f(x+p) = f(x)
for all x in its domain. If f(x) is periodic, its *period* is the smallest such number p. Graphically, this means that over adjacent intervals of length p, the graph of the function looks the same. The function completes one *cycle* over an interval of length equal to its period. If we know what one cycle of the graph looks like, then you know what the graph looks like over the entire domain of the function. The sine function is periodic, see the graphs of y = sint over [0, 2π] (Figure 6.21 a) and over the entire real line (Figure 6.21 b). Also, the period is not always the same, compare the graphs of one cycle of y = sint and the other sine function shown in Figure 6.21 c. The other function completes its cycle over the interval [0, π] so its period is π. We will show you how to find the period of the general trigonometric function later in this section.

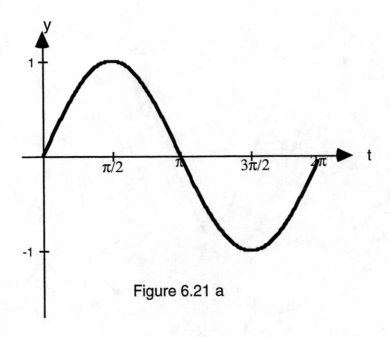

Figure 6.21 a

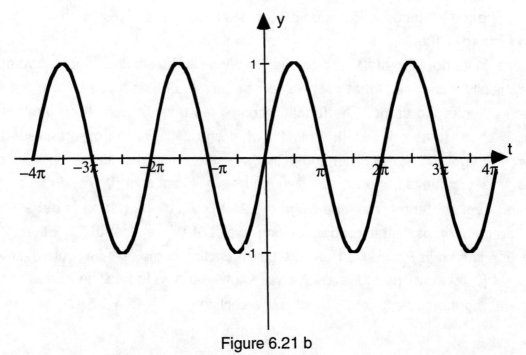

Figure 6.21 b

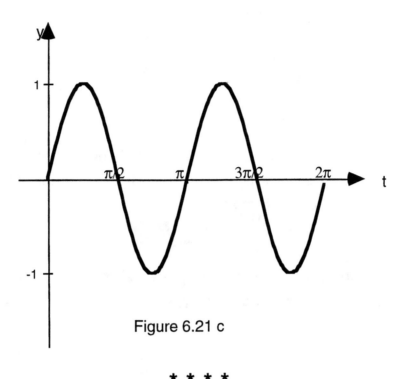

Figure 6.21 c

* * * *

Consider the general form for a sine function

$$y = a + b\sin(ct + d).$$

The vertical shift is a and the amplitude is |b|; these are the easy parts. To determine the period and shifts for this function, a little work is required. We compare one cycle of this function to one cycle of $y = \sin t$. The function $y = \sin t$ completes one cycle over the interval $[0, 2\pi]$. Thus, $\sin(ct + d)$ completes one cycle when $ct + d$ ranges from 0 to 2π. The cycle begins when

$$ct + d = 0$$
$$t = -\frac{d}{c}.$$

From this we see that the phase shift is $-\frac{d}{c}$. The cycle ends when

$$ct + d = 2\pi$$
$$t = \frac{2\pi}{c} - \frac{d}{c}.$$

Therefore, the period of $\sin(ct + d)$ is the length of the interval from the starting point, $t = -\frac{d}{c}$, to the ending point, $t = \frac{2\pi}{c} - \frac{d}{c}$. This is

$$\left|\left(\frac{2\pi}{c} - \frac{d}{c}\right) - \left(-\frac{d}{c}\right)\right| = \frac{2\pi}{|d|}.$$

Summarizing, $y = a + b\sin(ct + d)$ has

1) amplitude = |b|,

2) vertical shift = a,

3) horizontal shift = $-\dfrac{d}{c}$,

4) period = $\dfrac{2\pi}{|d|}$.

However, you do not need to memorize these four things, you will have a better understanding of the trigonometric functions if you perform the above analysis each time you graph one. Here are some examples.

Example 6.14

Determine the amplitude, period, phase shift and vertical shift of the function $y = 3 + 2\sin(2x - \pi)$. Graph one cycle.

Solution

Comparing

$$y = 3 + 2\sin(2x - \pi)$$

with the general form

$$y = a + b\sin(ct+d),$$

we see that a = 3, b = 2, c = 2 and d = -π. Thus the amplitude is 2 and the vertical shift is 3 units upward.

To determine the phase shift and period, we observe that $\sin(2x - \pi)$ starts its cycle when

$$2x - \pi = 0,$$

or when

$$x = \frac{\pi}{2}.$$

Thus $\frac{\pi}{2}$ is the phase shift. Continuing, $\sin(2x - \pi)$ ends its cycle when

$$2x - \pi = 2\pi,$$

or when

$$x = \frac{3\pi}{2}.$$

Hence the period is

$$\frac{3\pi}{2} - \frac{\pi}{2} = \pi.$$

The graph is shown in Figure 6.22.

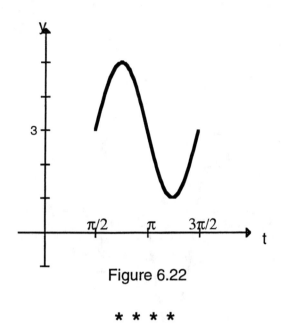

Figure 6.22

* * * *

Example 6.15

Determine the amplitude, period, phase shift, and vertical shift of the function $y = -1 + 5\sin(3t + 3\pi)$. Graph one cycle.

Solution

Comparing

$$y = -1 + 5\sin(3t + 3\pi)$$

with the general form

$$y = a + b\sin(ct - d),$$

we see that $a = -1$, $b = 5$, $c = 3$ and $d = 3\pi$. Thus the amplitude is 5 and the vertical shift is -1, or 1 unit downward.

To determine the phase shift and period, we observe that $\sin(3t + 3\pi)$ starts its cycle when

$$3t + 3\pi = 0,$$

or when

$$t = -\pi.$$

Thus $-\pi$ is the phase shift. Continuing, $\sin(3t + 3\pi)$ ends its cycle when

$$3t + 3\pi = 2\pi,$$

or when

$$t = -\frac{\pi}{3}.$$

Hence the period is

$$-\frac{\pi}{3} - (-\pi) = \frac{2\pi}{3}.$$

The graph is shown in Figure 6.23.

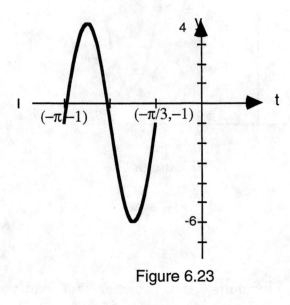

Figure 6.23

* * * *

In the next example, we show how to derive an equation for a sine function which has certain properties.

Example 6.16

Write the equation which defines the sine function which has vertical shift 1, phase shift $\pi/2$, period 6π and a maximum point at $(2\pi, 2)$. Graph one cycle.

Solution

We will use the general form $y = a + b\sin(ct - d)$. First we note that the vertical shift is 1, so $a = 1$. Since the phase shift is $\pi/2$ and the period is 6π, one cycle begins at $\pi/2$ and ends at $6\pi + \dfrac{\pi}{2} = \dfrac{13\pi}{2}$. Using this information, we see that

$$ct + d = 0 \text{ when } t = \frac{\pi}{2}$$

and

$$ct + d = 2\pi \text{ when } t = \frac{13\pi}{2}.$$

This gives us the two equations

$$\left(\frac{\pi}{2}\right)c + d = 0$$

$$\left(\frac{13\pi}{2}\right)c + d = 2\pi.$$

Now the constants c and d can be determined by simultaneously solving these two equations. The first one tells us that $d = -(\pi/2)c$; substitution of this into the second equation gives

$$\left(\frac{13\pi}{2}\right)c + d = 2\pi,$$

or

$$c = \frac{1}{3}.$$

Then we get $d = -\left(\frac{\pi}{2}\right)\left(\frac{1}{3}\right) = -\frac{\pi}{6}.$

We still need to find the amplitude and the coefficient b; using the given facts that the maximum value is 2 and the vertical shift is 1, it follows that the amplitude is 2 - 1 = 1 = |b|. But we still do not know if b is positive or negative. Now use the fact that the maximum occurs when $t = 2\pi$ to sketch one cycle of a graph of the function. See Figure 6.24. From this we see that the value of b in the equation must be positive and the desired equation is

$$y = 1 + \sin\left(\frac{1}{3}x - \frac{\pi}{6}\right).$$

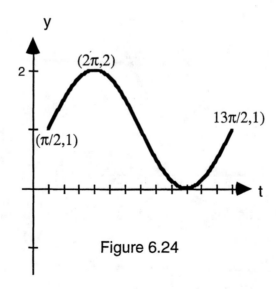

Figure 6.24

* * * *

Exercises

1.   Find a sine function which has one cycle the graph shown.

   a.

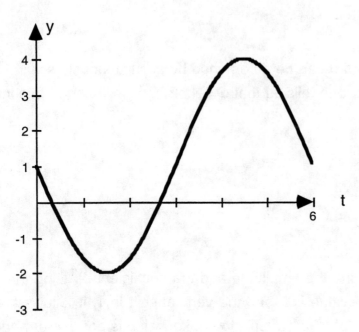

   b.

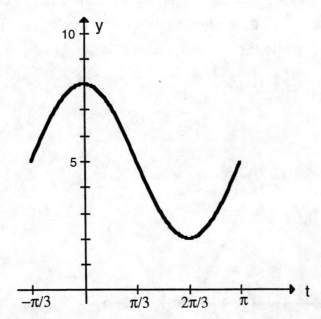

2.   Find the amplitude, period, and phase shift of the given function. Graph two cycles.

   a. $y = 3\sin\left(2t + \dfrac{\pi}{2}\right)$

   b. $y = 2\sin(3t - 12)$

c. $y = 1 + \sin\left(2t - \dfrac{\pi}{3}\right)$

d. $y = 5 - 3\sin\left[\left(\dfrac{\pi}{4}\right)t + \left(\dfrac{\pi}{2}\right)\right]$

e. $y = 2\sin(6t) - 3$

f. $y = 4\sin(2t - \pi)$

g. $y = \sin(t + 2) + 1$

3.   For each part below, write the equation which defines a sine function which has the given properties.

   a. Vertical shift 2, phase shift 1, period 6, a minimum point at (5.5, -4)
   b. Vertical shift 0, amplitude 2, phase shift 0, period 4π, maximum at π
   c. Vertical shift 1, amplitude 3, phase shift -π/4, period 2π, minimum at

π/4

   d. Vertical shift -2, amplitude 1/2, phase shift 3, period 2, minimum at 4.5

4.   Average daily high temperatures in a certain city are generally cyclic with a period of 12 months. On the average, the warmest day of the year is August 1 with a high temperature of 88°F and after that, the temperature cools down until it reaches the coldest high temperature of 28° F which occurs on January 31. Then, again on the average, temperatures warm until August 1. This pattern fairly closely repeats itself on a yearly cycle. Use this information to write the sine function which gives the average daily high temperature in this city as a function of day of the year. Let t = 0 represent January 1, t = 2 represent January 2, etc., and assume that there are 365 days in a year.

* * * *

6.8   The Cosine Function

   We can define the cosine function in much the same way that we defined the sine function. Let t be the radian measure of an angle in standard position and let (x,y) be the point of intersection of the terminal side of the angle with a circle of radius r > 0 and center at the origin. The cosine, abbreviated cos, is defined by

$$\cos t = \dfrac{x}{r}.$$

(See Figure 6.25.)

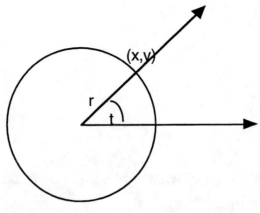

Figure 6.25

This definition is also independent of the size of the circle (why?), so that when the circle has radius 1 (the unit circle) then we have $\cos t = x$.

As with the sine, to see the graph of the cosine function, it is helpful to visualize the radius rotating counter clockwise once around the unit circle from its initial position. As the radius rotates, observe the values of $x = \cos t$. See Table 6.3.

As t varies from	cos t varies from
0 to π/2	1 to 0
π/2 to π	0 to -1
π to 3 π/2	-1 to 0
3π/2 to 2π	0 to 1

Table 6.3

On a different coordinate system with horizontal axis denoted t and vertical axis y, the graph of y = cos t over the interval [0, 2π] is shown. See Figure 6.26.

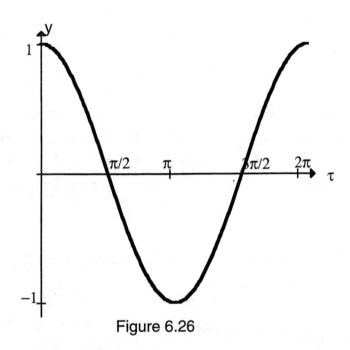

Figure 6.26

If the terminal side continues to rotate in either a clockwise or counter clockwise direction around the circle, it should be clear that the values of the cosine simply repeat over its entire domain (the real line). A portion of the graph of y = cos t is as shown in Figure 6.27.

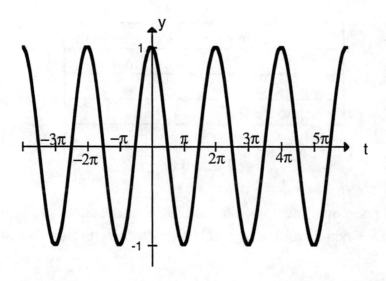

Figure 6.27

The domain of the cosine function consists of all real numbers. (Why?) The maximum value is 1, the minimum value is -1, so the range of the cosine function is the closed interval [-1, 1]. The cosine function is also periodic, i.e., its values repeat over certain adjacent intervals of equal length. The length of the smallest such interval is the period, and is the same as that of the sine, 2π.

Class Work

Use your calculator or computer to graph each of the following functions. Show at least two complete cycles.

1. $y = 2 \cos t$
2. $y = \cos 3t$
3. $y = \cos (t + 4)$
4. $y = 5 + \cos t$
5. $y = 5 + 2\cos(3t + 12)$

How would you change the graph of $y = \cos t$ to get each of the above graphs?

★ ★ ★ ★

6.9   Graphing the cosine function

The important features of the graph of the cosine function are the same as those for the sine function: vertical shift; amplitude; phase shift; and period. We consider cosine functions of the form

$$y = a + b\cos(ct + d)$$

and determine the important information in just the same manner as we did for the sine. Here are some examples.

Example 6.17

State the amplitude, period, vertical shift and phase shift of the function $y = 1 + 2\cos\left(3t + \dfrac{\pi}{2}\right)$. Graph one cycle.

Solution

Comparing to $y = a + b\cos(ct + d)$, we see that $a = 1$, $b = 2$, $c = 3$ and $d = \dfrac{\pi}{2}$. The graph is a cosine curve with vertical shift $a = 1$ and amplitude $|b| = 2$. One cycle of the cosine curve begins at

$$3x + \frac{\pi}{2} = 0 \text{ or at } x = -\frac{\pi}{6} \text{ (this is the phase shift),}$$

and ends at

$$3x + \frac{\pi}{2} = 2\pi \text{ or at } x = \frac{\pi}{2}.$$

Finally, we see that the period is

$$\left|\frac{\pi}{2} - \left(-\frac{\pi}{6}\right)\right| = \frac{2\pi}{3}.$$

The graph of this function is shown in Figure 6.28.

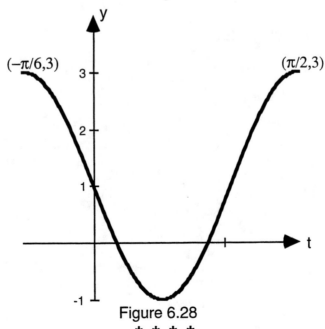

Figure 6.28

★ ★ ★ ★

Example 6.18

State the amplitude, period, vertical shift and phase shift of the function $y = -2 + 3\cos\left(2t + \dfrac{\pi}{2}\right)$. Graph one cycle.

Solution

Comparing
$$y = -2 + 3\cos\left(2t + \dfrac{\pi}{2}\right)$$

to the general form
$$y = a + b\cos(ct - d),$$

we see that a = -2, b = 3, c = 2 and $d = -\dfrac{\pi}{2}$. The graph is a cosine curve with vertical shift a = -2 and amplitude |b| = 3. One cycle of the cosine curve begins when

$$2t + \dfrac{\pi}{2} = 0 \text{ or } t = -\dfrac{\pi}{4}$$

and ends when

$$2t + \dfrac{\pi}{2} = 2\pi \text{ or } t = \dfrac{3\pi}{4}.$$

Hence the phase shift is -π/4 and the period is

$$\left|\dfrac{3\pi}{4} - \left(-\dfrac{\pi}{4}\right)\right| = \pi$$

Its graph is shown in Figure 6.29.

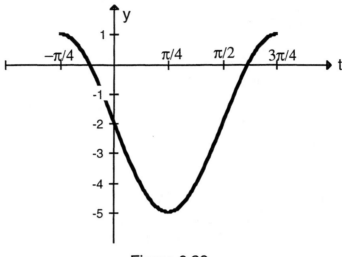

Figure 6.29

* * * *

In the following example, we illustrate how to write the equations of cosine functions which satisfy certain given properties.

Example 6.19

Write the cosine function which has vertical shift 5, phase shift -2, period 8, and a minimum point at (2,2).

Solution

Using the general form $y = a + b\cos(ct + d)$, we immediately get $a = 5$ and $d = 2$. Since the phase shift is -2, one cycle of this function begins at $t = -2$, i. e.,

$ct + d = 0$ when $t = -2$.

This gives the equation

* $-2c + d = 0$.

Since the period is 8, the cycle ends when $t = -2 + 8 = 6$, i. e.,

$ct + d = 2\pi$ when $t = 6$

which gives the second equation

* $6c + d = 2\pi$.

Now we solve the two equations * simultaneously. From the first, we get $d = 2c$, and then substitute this into the second to get

$6c + (2c) = 2\pi$.

From this,

$$c = \frac{\pi}{4} \text{ and } d = \frac{\pi}{2}.$$

Finally, we need to find the amplitude and the constant b. Since the vertical shift is 5 and the minimum value of the function is 2, the amplitude is 5 - 2 = 3. To see if b is positive or negative, we observe that the minimum occurs at t = 2, so b > 0, i. e., b = 3. Hence the desired equation is

$$y = 5 + 3\cos\left[\left(\frac{\pi}{4}\right)t + \left(\frac{\pi}{2}\right)\right],$$

and one cycle of the graph is shown in Figure 6.30.

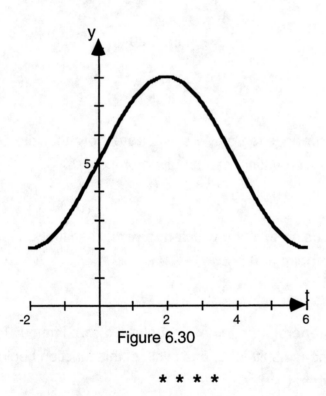

Figure 6.30

* * * *

Exercises

1.   Find a cosine function with one cycle of the graph shown.

a.

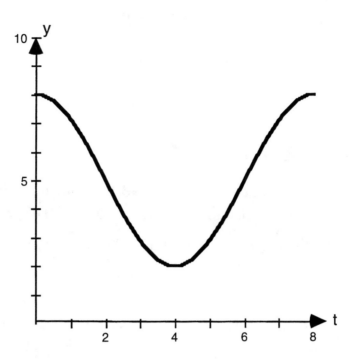

b.

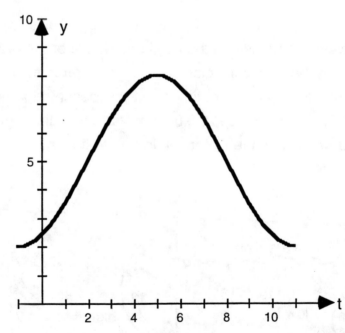

2.   Find the amplitude period and phase shift of the given function. Graph one complete cycle.

a. $y = 2\cos\left(3t - \dfrac{\pi}{2}\right)$

b. $y = \left(\dfrac{4}{3}\right)\cos\left(2t + \dfrac{\pi}{2}\right)$

c. $y = 1 + \cos\left(2t + \dfrac{\pi}{2}\right)$

d. $y = 2 - 3\cos\left[\left(\dfrac{\pi}{3}\right)t + \pi\right]$

3. Write the equation which defines each cosine function which has the given properties.

    a. vertical shift -2, phase shift $-\dfrac{\pi}{3}$, period $\dfrac{4\pi}{3}$, maximum point $\left(-\dfrac{\pi}{3}, 2\right)$

    b. Vertical shift 2, phase shift 1, period 6, a minimum point at (4, -4)

    c. Vertical shift 0, amplitude 2, phase shift 0, period 4π, maximum at 4π

    d. Vertical shift 1, amplitude 3, phase shift $-\dfrac{\pi}{4}$, period 2π, minimum at $-\dfrac{\pi}{4}$

    e. Vertical shift -2, amplitude $\dfrac{1}{2}$, phase shift 3, period 2, minimum at 4.

4. In the example at the beginning of this chapter in which we introduced the sine function, we measured the height of the water. Now consider a function which measures the horizontal distance from the water line on the beach at mid-tide to the water line at different times. On this beach, high-tide occurs at noon, mid-tide at 3:00 p.m.; low-tide at 6:00 p.m., and high-tide again at midnight. At high-tide and at low-tide the water line is 100 feet away from the mid-tide point. (See the Figure below.)

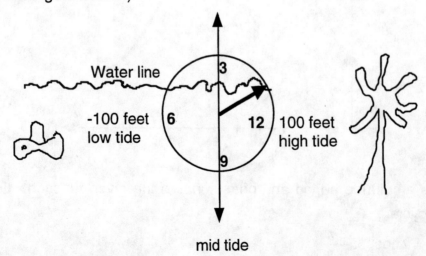

mid tide

This information is summarized in the table below using t = 0 for noon and x = 100 for the water line at high-tide.

t	x
0	100
3	0
6	-100
12	100

As before, these data can be conveniently represented by a clock with one hand of length 100 feet placed on an (x,y) coordinate system as shown below.

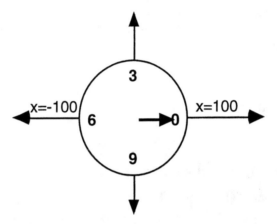

At time t = 0 the hand is on the positive x axis and rotates counterclockwise one complete revolution is 12 hours. The tip of the hand is always at the water line. As the clock hand turns, it determines points on a coordinate system with the first coordinate x indicating the horizontal distance of the waterline from the mid-tide mark. The tip of the hand traces the graph of a cosine function; write the equation of this function and sketch one cycle of its graph.

* * * *

6.10 The Tangent Function

The next trigonometric function we study is the tangent function. In order to define this function, we again let t be the radian measure of an angle in standard position and let (x,y) be the point of intersection of the terminal side of the angle with a circle of radius r > 0 which has center at the origin. (See Figure 6.31.)

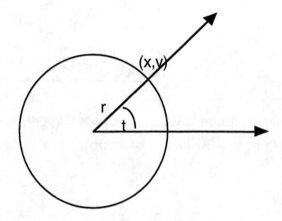

Figure 6.31

The *tangent*, abbreviated tan, is defined by

$$\tan t = \frac{y}{x}.$$

As with sine and cosine, this definition is also independent of the length of the radius of the circle. (Why?)

From the definition of tangent, we see that its domain consists of all t for which $x \neq 0$; this is the set

$$\left\{ t \middle| t \neq \pm\frac{\pi}{2}, \pm\frac{3\pi}{2}, \pm\frac{5\pi}{2}, \ldots \right\}.$$

To see the graph of the tangent, it is helpful to visualize the radius of the circle as the terminal side of the angle rotating first counterclockwise from the positive x-axis to $\pi/2$, then clockwise from the positive x-axis to $-\pi/2$. As the radius rotates, observe the values of $\tan(t) = \frac{y}{x}$ (see Table 6.2).

As t varies from	tan t varies from
0 to $\pi/2$	0 to +infinity
0 to $-\pi/2$	0 to -infinity

Table 6.2

Class Work

Complete the table as the terminal side of the angle rotates first clockwise from the negative x-axis (π) to $\dfrac{\pi}{2}$ then counterclockwise from the negative x-axis to $\dfrac{3\pi}{2}$. You should see that the values of tan t from $\dfrac{\pi}{2}$ to $\dfrac{3\pi}{2}$ repeat the values of tan t from $-\dfrac{\pi}{2}$ to $\dfrac{\pi}{2}$.

* * * *

From Table 6.2, it is easy to see that the range of tangent consists of all real numbers. On a different coordinate system with horizontal axis denoted t and vertical axis y, the graph of y = tan t over the interval $\left(-\dfrac{\pi}{2}, +\dfrac{\pi}{2}\right)$ is shown. See Figure 6.32.

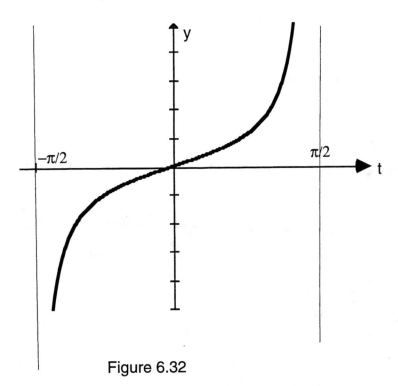

Figure 6.32

If the terminal side continues to rotate in either a clockwise or counter clockwise direction around the circle, it should be clear that the values of the tangent simply repeat over intervals of length π over the real line. Therefore the period of tangent is π. A portion of the graph of y = tan t is as shown in Figure 6.33.

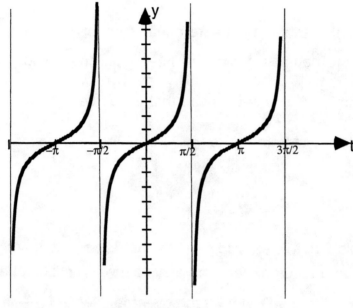

Figure 6.33

You can see from the graph that one cycle is completed for $-\dfrac{\pi}{2} < t < +\dfrac{\pi}{2}$.

Vertical asymptotes occur at

$$t = \ldots, -\dfrac{3\pi}{2}, -\dfrac{\pi}{2}, \dfrac{\pi}{2}, \dfrac{3\pi}{2}, \ldots$$

Example 6.20

   Determine the period, phase shift, and then graph one cycle of

$$y = \tan\!\left(2t - \dfrac{\pi}{2}\right).$$

Solution

   One cycle will begin when

$$2t - \dfrac{\pi}{2} = -\dfrac{\pi}{2},$$

or when

$$t = 0;$$

it will end when

$$2t - \dfrac{\pi}{2} = -\dfrac{\pi}{2},$$

or when

$$t = \dfrac{\pi}{2}.$$

Hence the period is

$$\frac{\pi}{2} - 0 = \frac{\pi}{2}.$$

To find the phase shift, we solve for t when

$$2t - \frac{\pi}{2} = 0$$

to get $t = \frac{\pi}{4}$; this is the phase shift.

The graph of $y = \tan\left(2t - \frac{\pi}{2}\right)$ will look almost the same as the graph of $y = \tan t$

except it will be shifted horizontally to the right $\frac{\pi}{2}$ units and have the shorter

period $\frac{\pi}{2}$. See Figure 6.34.

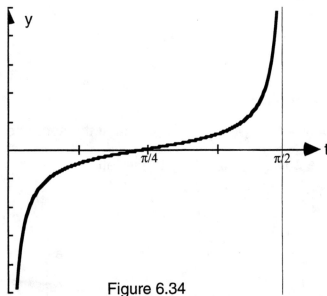

Figure 6.34

* * * *

Example 6.21

Determine the period, phase shift, and then graph one cycle of

$$y = 1 + \tan\left(\frac{t}{3} + \pi\right).$$

Solution

One cycle will begin when

$$\frac{t}{3} + \pi = -\frac{\pi}{2},$$

or when

$$t = -\frac{9\pi}{2};$$

it will end when

$$\frac{t}{3} + \pi = \frac{\pi}{2},$$

or when

$$t = -\frac{3\pi}{2}.$$

Hence the period is

$$-\frac{3\pi}{2} - \left(-\frac{9\pi}{2}\right) = 3\pi.$$

To find the phase shift, we solve for t when

$$\frac{t}{3} + \pi = 0$$

to get t = -3π ; this is the phase shift.

The graph of $y = \tan\left(\frac{t}{3} + \pi\right)$ will look almost the same as the graph of y = tan t except it will be shifted horizontally to the left -3π units, shifted vertically 1 unit, and have period 3π. See Figure 6.35.

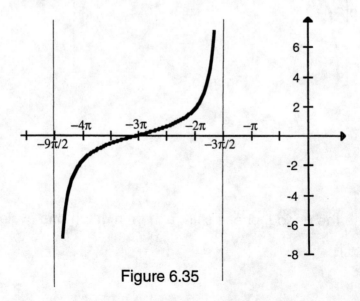

Figure 6.35

* * * *

Class Work

Determine the period and phase shift of each function.

1. $y = \tan(t - \pi)$.

2. $y = -4\tan\left(\dfrac{t}{2}\right)$.

$$\bigstar \; \bigstar \; \bigstar \; \bigstar$$

Exercises

In problems 1 - 4 match the given function to one of the graphs a - d.

1. $y = -\tan\left(t + \dfrac{\pi}{2}\right)$ 2. $y = \tan\left(\dfrac{t}{2} - \pi\right)$

3. $y = 3 + \tan 2t$ 4. $y = 3\tan\left(2t + \dfrac{\pi}{4}\right)$

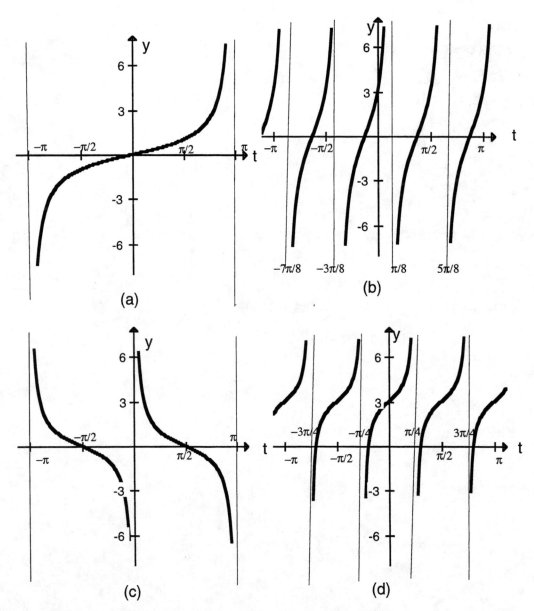

(a)

(b)

(c)

(d)

In problems 5 - 8, determine the period, phase shift and graph one cycle of each function.

5. $y = 2 + \tan\left(\dfrac{t}{4}\right)$ 6. $y = -\tan(3t)$

7. $y = \tan\left(t + \dfrac{\pi}{2}\right)$ 8. $y = 1 - \tan\left(\dfrac{t}{2} + \pi\right)$

∗ ∗ ∗ ∗

6.11 The Secant, Cosecant, and Cotangent Functions

In this section, we examine the other three trigonometric functions. The most often used are the ones already studied, sine, cosine, and tangent. The secant, cosecant and cotangent can be expresses in terms of the other functions so this section may be optional.

6.11.1   The secant Function

Once again let t be the radian measure of an angle and let (x,y) be the point of intersection of the terminal side of the angle with a circle of radius r > 0. The *secant*, abbreviated sec, is defined by

$$\sec t = \frac{r}{x}$$

(Again, notice that this definition is independent of the radius of the circle.) It is important to note that $\sec(t) = \dfrac{1}{\cos(t)}$ except for values of t where cos t = 0.

Thus the domain consists of all values of t for which cos t = x/r ≠ 0; i. e., x ≠ 0. From the graph of cosine, we see that cos t = 0 for t = ±π/2, ±3π/2, ±5π/2, ...; therefore the domain of secant is the set

{t I t ≠ ±π/2, ±3π/2, ±5π/2, ...}.

The range is the set of numbers {y I y ≤ -1 or y ≥ 1}, i. e., (−∞,−1]∪[1,∞). Since $\sec(t) = \dfrac{1}{\cos(t)}$, the graph of the secant function can be obtained from observation of the graph of cosine. The secant graph unfolds as we consider the values of t as it varies from 0 to 2π. See the Table 6.5 below.

As t goes from	cos t goes from	and sec t goes from
0 to π/2	1 to 0	1 to +∞
π/2 to π	0 to -1	-∞ to -1
π to 3π/2	-1 to 0	-1 to -∞
3π/2 to 2π	0 to 1	∞ to 1

Table 6.5

Figure 6.36a shows one cycle of y = cos t with one cycle of y = sec t sketched around it. One complete cycle of the graph of the secant function is shown in Figure 6.36b.

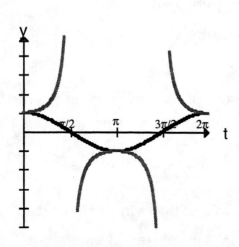

Figure 6.36a   Figure 6.36b

Just like the cosine function, the secant function is periodic with period 2π. Vertical asymptotes occur at

$$t = ..., -\frac{3\pi}{2}, -\frac{\pi}{2}, \frac{\pi}{2}, \frac{3\pi}{2}, ...$$

Amplitude is undefined for secant, but vertical and phase shifts are relevant and can be determined in much the same way as for the other trigonometric functions. The examples below illustrate these techniques.

Example 6.22

Determine the period and vertical and phase shifts, then and graph one cycle of

$$y = \sec\left(t + \frac{\pi}{2}\right).$$

Solution

One cycle of the secant will begin when

$$t + \frac{\pi}{2} = 0,$$

or when

$$t = -\frac{\pi}{2}.$$

It will end when

$$t + \frac{\pi}{2} = 2\pi$$

or when

$$t = 2\pi - \frac{\pi}{2} = \frac{3\pi}{2}.$$

Hence the phase shift is $-\pi/2$ and the period is

$$\frac{3\pi}{2} - \left(-\frac{\pi}{2}\right) = 2\pi.$$

The vertical shift is 0. The graph will look like the graph of $y = \sec t$ shifted horizontally $\frac{\pi}{2}$ units to the left. See Figure 6.37.

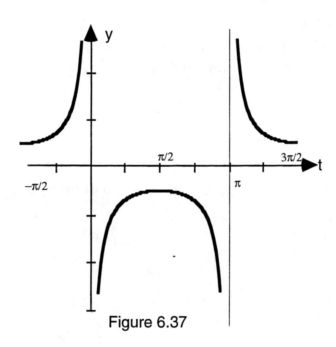

Figure 6.37

* * * *

Example 6.23

Determine the period, vertical shift and phase shift, then graph one cycle of $y = 1 + 2\sec\left(\frac{t}{2} - \pi\right)$.

Solution

One cycle of the secant will begin when

$$\frac{t}{2} - \pi = 0,$$

or when

$$t = 2\pi.$$

It will end when

$$\frac{t}{2} - \pi = 2\pi,$$

or when

$$\frac{t}{2} = 3\pi$$

so that

$$t = 6\pi.$$

Thus, the phase shift is 2π and the period is

$$6\pi - 2\pi = 4\pi.$$

The vertical shift is 1. The graph will look like the graph of y = 2sec x shifted horizontally 2π units to the right and 1 unit up.

* * * *

Class Work

Find the phase shift, vertical shift and period of each function.

1. $y = \sec(t - \pi)$
2. $y = 1 - \sec(3t + \pi)$
3. $y = -3\sec t$

* * * *

Exercises

In exercises 1-4, match each function to one of the graphs A-D.

1. $y = 3\sec(4t)$ 2. $y = -\sec(t + \pi)$

3. $y = 3 + \sec(4t - \dfrac{\pi}{2})$ 4. $y = 1 - \sec(2t - \pi)$

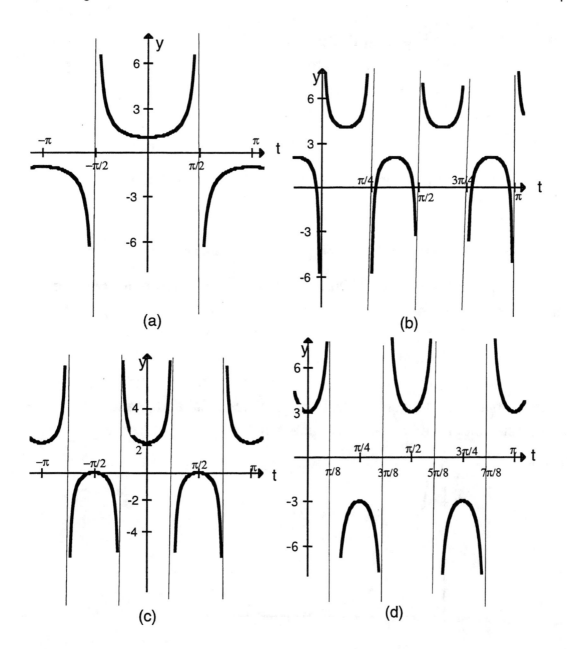

(a)

(b)

(c)

(d)

Graph each function showing at least one cycle. Indicate period, phase shift, and vertical shift.

5. $y = -\sec t$

6. $y = 2\sec(3t)$

7. $y = 3 + \sec\left(t + \dfrac{\pi}{4}\right)$

8. $y = \sec(2t - \pi)$

★ ★ ★ ★

6.11.2   The Cotangent and Cosecant Functions

One last time, let t be the radian measure of an angle and let (x,y) be the point of intersection of the terminal side of the angle with a circle of radius r > 0. The *cotangent,* abbreviated cot, is defined by

$$\cot(t) = \frac{x}{y}.$$

Note that the cotangent function is the reciprocal of the tangent, i. e.,

$$\cot(t) = \frac{1}{\tan(t)},$$

except for t = any multiple of π/2. When t = an odd multiple of $\frac{\pi}{2}$, the tangent is undefined but the cotangent is 0. On the other hand, at even multiples of π/2, the tangent is 0 and the cotangent is undefined. Hence the domain of the cotangent is the set {t | t ≠ 0, ±π, ±2π, ±3π, . . . }, and its graph can be seen by consideration of the fact that cot t = 1/tan t. One cycle is shown in Figure 6.38. Note that vertical asymptotes occur at t = 0, ±π, ±2π, ±3π, . . . , and the period is the same as that of the tangent which is π. Again, vertical and phase shifts and period are relevant properties but amplitude is not.

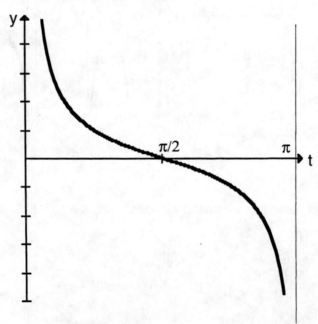

Figure 6.38

* * * *

The *cosecant* function, abbreviated csc, is defined to be

$$\csc(t) = \frac{r}{y}$$

and we note that the cosecant is the reciprocal of the sine, i. e., $\csc(t) = \dfrac{1}{\sin(t)}$ except for values of t where sin t = 0. Thus, the domain of the cosecant is the set {t I t ≠ 0, ±π, ±2π, ±3π, . . . }. Its graph can be seen by consideration of the fact that csc t = 1/sin t. One cycle is shown in Figure 6.39. Note that vertical asymptotes occur at t = 0, ±π, ±2π, ±3π, . . . , and the period is 2π. And, once again, vertical and phase shifts and period are relevant properties but amplitude is not.

Techniques for graphing cotangent and cosecant functions are the same as those used for graphing the other trigonometric functions.

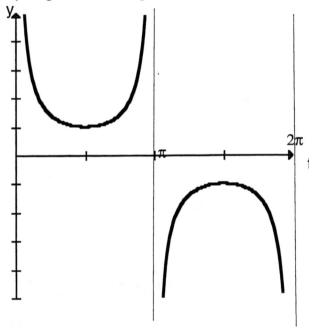

Figure 6.39

Class Work

1. Determine the period and phase shift of the function

   y = cot (2t - π).

2. Determine the period, phase shift and vertical shift of the function

   $$y = 1 + \csc\left(t - \frac{\pi}{2}\right).$$

★ ★ ★ ★

Exercises

In exercises 1 - 3, determine the period and phase shift of each function. Graph one cycle.

1. $y = \cot(t + 3\pi)$

2. $y = -\cot(3t)$

3. $y = 4\cot\left(2t - \dfrac{\pi}{2}\right)$

In exercises 4 - 6, determine the period, phase shift and vertical shift of each function. Graph one cycle.

4. $y = \csc\left(t + \dfrac{\pi}{3}\right)$

5. $y = 2 + \csc(t - \pi)$

6. $y = -2\csc(6t + 3\pi)$

* * * *

CHAPTER SEVEN
SNOW-FED RIVERS: TEMPERATURE AND PRECIPITATION

Introduction

In this and in the next chapter, we return to the Village of River City (Chapter 4) which is dependent on the Nizhoni River for its water supply. Due to its growing population, its residents are concerned about the future availability of water from the river; this necessitates a determination of the monthly and annual amount of water which flows through the river.

The forecasting of annual water supply is based primarily on the amount of precipitation, particularly in the form of snow, and temperatures in the region. Average monthly temperatures and precipitation generally are cyclic on an annual basis, and of course, snowfall is dependent upon both these factors. The patterns for our region are: most of the precipitation comes in winter and spring, tapers off to a low point in fall and then increases after that; temperatures follow the known pattern, lowest in winter and highest in summer. For this study data describing temperature and precipitation will be provided. Based on these data we will derive functions which describe both of these, and together with a function which estimates the amount of snow melt at a particular temperature, use these to predict streamflow, the amount of water which goes down the river in some interval of time.

7.1   Predicting Streamflow

In the following sections we will derive functions which will be used to determine the streamflow for the Nizhoni River. The functions are all functions of the variable t, $0 \le t \le 12$, where t represents time (in months) from the beginning of the year.

$A(t)$ = temperature at time t (degrees Fahrenheit)

$R(t)$ = rate of precipitation at time t (inches per month)

$W(t)$ = rate of precipitation over the watershed at time t ($\times 10^6$ ft^3 per month)

$M(t)$ = rate of snow melt over the watershed at time t (inches per month)

$S(t)$ = rate of snow melt over the watershed at time t ($\times 10^6$ ft^3 per month)

Note that R(t) measures the rate of precipitation at a particular point while W(t) measures the rate of precipitation over the entire watershed, and similarly for

M(t) and S(t). From these we will derive a function F(t) which describes the streamflow of the Nizhoni River in million cubic feet per month.

* * * *

7.2   Temperature

Air temperature is a cyclic phenomenon with a cycle of 12 months. We derive a function A(t) to estimate the temperature at time t, based on an average taken over a number of years. Note that this function of course is not a very accurate indicator of temperature at a specific time since it is normally cool in the morning, warmer during the day, and cooler again at night. Also, for any particular year and at different times of day the temperature will vary from the average. However, it will allow us to approximate average temperature for certain times. For example, suppose A (2.5) = 27. This means that at time 2.5 (some time on March 16th) the temperature will be approximately 27 degrees.

Group Work

The Nizhoni River originates high in the mountains and even during the warmest times of year the temperature is quite low. Based on average temperature data, we make the following assumptions:

the minimum temperature of 10.0° occurs at the end of January (t = 1);
the maximum temperature of 49.7° occurs at the end of July (t = 7);
temperature is cyclic on an annual basis.

Determine a function which describes the average daily temperature at time t; use a sine function of the form A(t) = a + bcos(ct + d) which satisfies all of these assumptions. Follow the steps below (round coefficients to two decimal places).

1.   Sketch a rough graph of how the function should look. Label the maximum and minimum points.

2.   What is the period?

3.   Determine the amplitude. Is b positive or negative?

4.   Determine the phase shift and the vertical shift.

5. Write the function A(t) satisfying the properties in the previous steps, and graph one cycle of the function representing one calendar year.

Use the function A(t) to answer the remaining questions about temperature in this chapter.

6. What is the temperature at the end March? When does the temperature again reach that level?

7. When will the temperature be 32°?

8. We assume that when the temperature is above 32°, precipitation means rain; otherwise precipitation is in the form of snow and it sticks. Determine the period during which temperatures are above freezing, i.e., when precipitation is in the form of rain and the snow melts. (Round your answers to one decimal place.)

9. Indicate on the graph the times of the year when the temperature is 32°.

* * * *

TEMPERATURE SUMMARY

Fill in the required information and save it on a copy of this page; you will need
this later. When writing in your answers be sure to include the correct units; this
is important and you may forget them later on when you need them.

TEMPERATURE FUNCTION (See question #5.)

INTERVAL OF TIME WHEN TEMPERATURE IS ABOVE FREEZING (RAIN AND
SNOW MELT) (See question #8.)

INTERVAL OF TIME WHEN TEMPERATURE IS FREEZING OR BELOW (SNOW
AND ICE) (See question #8.)

7.3   Inverse Trigonometric functions

Since the trigonometric functions are periodic, they are not one-to-one. So before we can have any discussion about their inverses, they must be restricted to an interval over which they are one-to-one. In this section, we do exactly that for the three functions, sine, cosine, and tangent. The domain of each of these functions will be restricted to an interval chosen so that it is close to the origin, the restricted function is one-to-one, and its range is the same as the range of the unrestricted function.

7.3.1   The Inverse Sine

The closed interval $\left[-\dfrac{\pi}{2},\dfrac{\pi}{2}\right]$ is centered at the origin and, restricted to this interval, the sine function is one-to-one and has range [-1,1]. The graph is shown in Figure 7.1.

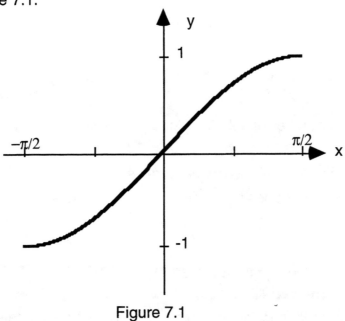

Figure 7.1

This restricted sine function will have an inverse and is defined by

$$y = \sin^{-1}x \text{ if } \sin y = x,\ -\frac{\pi}{2} \le y \le \frac{\pi}{2}.$$

Notice that this is the same as the restricted sine function with the role of x and y interchanged. The domain is [-1,1] and the range is $\left[-\dfrac{\pi}{2},\dfrac{\pi}{2}\right]$; its graph is shown in Figure 7.2.

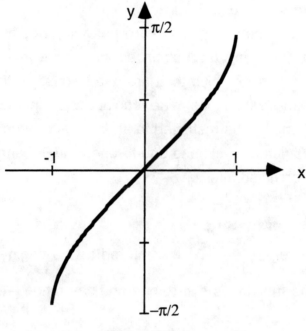

Figure 7.2

It is important to note that for any x, -1 $\leq$ x $\leq$ 1, the equation,

   sin y = x,

will have an infinite number of solutions for y whereas the equation,

   $\sin^{-1}x = y$,

has a unique solution for y. This is what makes the inverse sine a function of x, while the relation sin y = x does not define y as a function of x.

7.3.2   The Inverse Cosine

In this case we cannot restrict to any interval which is centered about the origin because the cosine function will not be one-to-one over such an interval. However, the interval [0,π] has its left end-point at x = 0 and, over this interval, the cosine function is one-to-one and has range [-1,1]. The graph is shown in Figure 7.3.

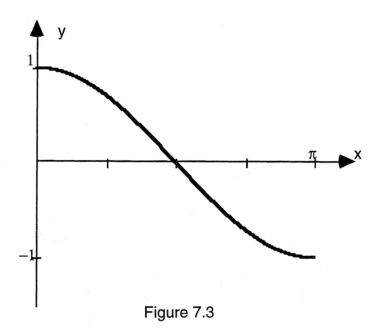

Figure 7.3

Hence we define the inverse cosine function by

$y = \cos^{-1}x$ if $\cos y = x$, $0 \le y \le \pi$.

Its domain is [-1,1] and its range is [0,π]; its graph is shown in Figure 7.4. Once

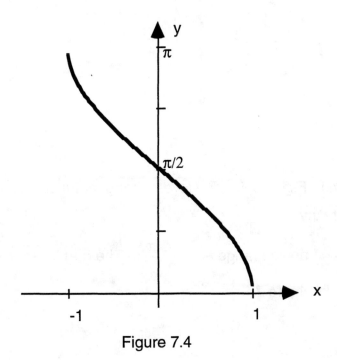

Figure 7.4

again, note that for any value of x, $-1 \le x \le 1$, the equation,

$\cos y = x$,

will have infinitely many solutions for y. However, there will be only one solution to

$$\cos^{-1}x = y.$$

$$* \ * \ * \ *$$

7.3.3   The Inverse Tangent

Here the natural restriction is the open interval $\left(-\dfrac{\pi}{2},\dfrac{\pi}{2}\right)$. The tangent function is one-to-one over this interval and has as its range all real numbers.

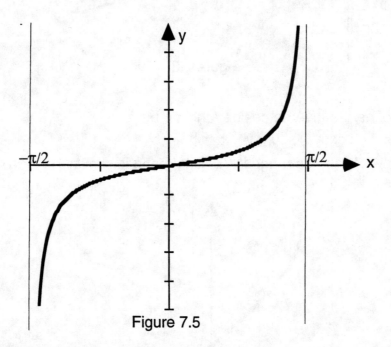

Figure 7.5

Its graph is shown in Figure 7.5. The inverse tangent function is defined by

$$y = \tan^{-1}x \text{ if } \tan y = x, \ -\frac{\pi}{2} < y < \frac{\pi}{2}.$$

Its domain is $(-\infty,\infty)$ and its range is $\left(-\dfrac{\pi}{2},\dfrac{\pi}{2}\right)$. The graph of the inverse tangent function is shown in Figure 7.6.

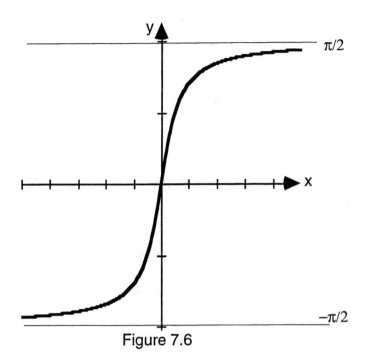

$\pi/2$

$-\pi/2$

Figure 7.6

How many solutions are there for y if tan y = x? How many if tan⁻¹x = y?

$$* \quad * \quad * \quad *$$

7.4   Reference Angles, Trigonometric Equations and Inequalities

Suppose we need to know all values of t for which

$$\sin(t) = \frac{1}{2}.$$

An immediate solution is

$$t = \sin^{-1}\left(\frac{1}{2}\right) = \frac{\pi}{6} \text{ (or, in decimal approximation, 0.5236).}$$

However, as pointed out in the previous section, the inverse sine gives only one answer for t, but there are an infinite number of such values for t. Since the sine function is periodic with period 2π, we only have to look in the interval $[0, 2\pi]$ for solutions and then translate these by multiples of 2π. To find solutions for t in this interval, we think of t as the radian measure of angles in standard position between 0 and 2π. If we consider the circle with center at the origin and radius 1, then the definition of sine gives us

$$\sin(t) = \frac{y}{1} = \frac{1}{2},$$

so t will represent measures of angles with terminal sides which intersect the circle at points with y-coordinates y = 1/2. Refer to Figure 7.7. There are exactly two of these, one in Quadrant I and one in Quadrant II. We already have one value for t, $t = \dfrac{\pi}{6}$. This is the measure of the angle with terminal side in Quadrant I which has sine 1/2. To find the angle with terminal side in Quadrant II, note that the measure of the acute angle between its terminal side and the negative x-axis is the same as the known measure of the angle in Quadrant I, i. e., π/6. Therefore the second solution in the interval [0, 2π] is

$$t = \pi - \frac{\pi}{6} = \frac{5\pi}{6}.$$

In this example, the angle π/6 is called the *reference angle* .

Finally, all solutions are obtained by translating these two by integral multiples of 2π,

$$t = \frac{\pi}{6} \pm 2\pi n, \; t = \frac{5\pi}{6} \pm 2\pi n.$$

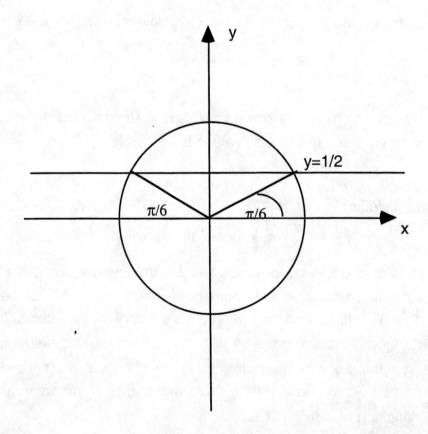

Figure 7.7

* * * *

The important thing about the above example is the reference angle. In general, if t is the measure of any angle in standard position with its terminal side not on a coordinate axis, then its *reference angle* ρ is the acute angle formed by its terminal side and the x-axis. See Figure 7.8.

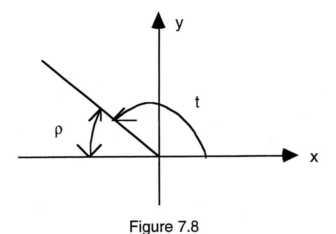

Figure 7.8

For an angle with measure t radians, the importance of its reference angle ρ is this: if F is any trigonometric function,

$$F(t) = \pm F(\rho),$$

where + or - is determined by the sign of the function F in the Quadrant where the terminal side of t lies. (The student should verify this by considering the definitions of the trigonometric functions.) With this information, we can use reference angles and the knowledge of the signs of the trigonometric functions in the various quadrants to find all solutions to trigonometric equations of the type seen above. Here are some more examples.

Example 7.1

Find all solutions over the interval [0, 2π] to the equation

$$\sin t = \frac{\sqrt{3}}{2} \quad .$$

Solution

One solution is given by

$$t = \sin^{-1}\left(\frac{\sqrt{3}}{2}\right) = \frac{\pi}{3} \quad \text{(or 1.0472 in decimal approximation);}$$

this solution is in Quadrant I. But since the sine is also positive in Quadrant II, there is also a solution there and its reference angle is $\pi/3$. Hence the other solution is

$$t = \pi - \frac{\pi}{3} = \frac{2\pi}{3}.$$

★ ★ ★ ★

Example 7.2

Find all solutions over the interval $[0, 2\pi]$ to the equation

$$\cos t = -0.3456.$$

Solution

One solution is given by

$$t = \cos^{-1}(-0.3456) = 1.9237.$$

Since $t = 1.9237$ is between $\pi/2$ and π, this angle lies in Quadrant II and its reference angle will be

$$\pi - 1.9237 = 1.2179.$$

To find other solutions, we note that cosine is also negative in Quadrant III, so there will be a another solution there. This solution will also have reference angle 1.2179, so the other solution is

$$t = \pi + 1.2179 = 4.3595.$$

★ ★ ★ ★

Example 7.3

Find all solutions over the interval $[0, 2\pi]$ to the equation

$$\tan t = 4.56.$$

Solution

One solution is given by

$$t = \tan^{-1}(4.56) = 1.3549.$$

This is in Quadrant I, but tangent is also positive in Quadrant III, so there will be another solution there with reference angle 1.3549. Hence this solution is

$$t = \pi + 1.3549 = 4.4965.$$

★ ★ ★ ★

Example 7.4

Find all solutions over the interval $[0, 2\pi]$ to the equation

$$\cos t = 0.2345.$$

Solution

One solution is given by

$$t = \cos^{-1}(0.2345) = 1.3341.$$

This is in Quadrant I, but cosine is also positive in Quadrant IV, so there will be another solution there with reference angle 1.3341. Hence this solution is

$$t = 2\pi - 1.3341 = 4.9491.$$

★ ★ ★ ★

Class Work

1. Find all solutions over the interval $[0, 2\pi]$ to the equation

   $$\sin t = 0.6543.$$

2. Find all solutions over the interval $[0, 2\pi]$ to the equation

   $$\tan t = -6.54.$$

3. Find *all solutions* to the equation

   $$\cos t = 0.4334.$$

★ ★ ★ ★

7.5   Trigonometric Equations

In Section 7.2, you solved a trigonometric equation to find out when the temperature is above freezing in the region around the Nizhoni River. In general, a trigonometric equation is an equation which involves one or more trigonometric function. The examples in the previous section are simple trigonometric equations; in this section we look some slightly more complicated ones. There are usually two ways to solve these equations, one is algebraically and the other is graphically. When solving algebraically, one first isolates the trigonometric function of the unknown and then uses the inverse of the function. The inverse function only gives one possible solution and there may be others, so it is important to pay attention to this. Solving graphically is usually easier and involves finding the intersection of graphs on the calculator. The idea is to graph the left and right sides of the equation as two separate functions; the solutions to the equation are the values of the independent variable for which the two functions are equal. These will be the x-coordinates of the points of intersection of the graphs.

It is important for you to understand the principles behind both methods so we illustrate each technique in the examples below.

Example 7.5

Reconsider Example 7.1 in the preceding section. We found all solutions over the interval [0, 2π] to the equation

$$sin\, t = \frac{\sqrt{3}}{2}\ .$$

We now show how this can be accomplished graphically. Enter

$$y_1 = sin\ t \text{ and } y_2 = \frac{\sqrt{3}}{2}$$

in your calculator; set the x interval from 0 to 2π and the y interval from -1 to 1 (why?) and graph. You see that there are two points of intersection; these can be found by using the "intersect" feature on your calculator to be

$$t = 1.0472\left(=\frac{\pi}{3}\right),\ t = 2.0944\left(=\frac{2\pi}{3}\right).\ \text{ See Figure 7.9 below.}$$

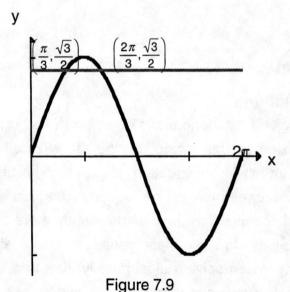

Figure 7.9

* * * *

Example 7.6

Solve the equation

$$2\sin x \cos x - \sin x = 0$$

for all x over the interval $[0, 2\pi]$.

Solution

We illustrate both the algebraic method and the graphical methods.

A) Algebraic

Factor out sin x to get

$$\sin x(2\cos x - 1) = 0.$$

Now set each factor = 0.

$$\sin x = 0, \qquad\qquad 2\cos x - 1 = 0$$

$$\cos x = 1/2$$

First, recall that over $[0, 2\pi]$,

$$\sin x = 0 \text{ if } x = 0, \pi \text{ or } 2\pi.$$

Looking at the second factor, we see that one solution is given by

$$x = \cos^{-1}\left(\frac{1}{2}\right) = \frac{\pi}{3} \text{ (or, in decimal approximation, 1.0472.}$$

Using $\pi/3$ as a reference angle, we see that another solution lies in Quadrant IV,

$$x = 2\pi - \frac{\pi}{3} = \frac{5\pi}{3}.$$

Hence there are five solutions to the original equation,

$$x = 0, \pi, 2\pi, \frac{\pi}{3} \text{ or } \frac{5\pi}{3}.$$

B) Graphical

Although it is not necessary, we rearrange the equation as

$$2\sin x \cos x = \sin x.$$

Now, on your calculator, enter $y_1 = 2\sin x \cos x$ and $y_2 = \sin x$, set the x interval from 0 to 2π and the y interval from -2 to 2 (why?), then graph. You see five points of intersection which can be found using "intersect" to be (of course)

$$x = 0, 3.1416 (= \pi), 6.2832 (= 2\pi), 1.0472\left(=\frac{\pi}{3}\right), \text{ or } 5.2360\left(=\frac{5\pi}{3}\right).$$

See Figure 7.10.

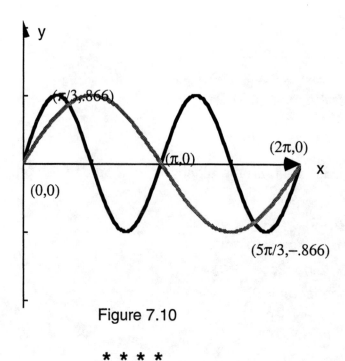

Figure 7.10

* * * *

Example 7.7

Find all solutions to the equation

3.2sin x + 1.7 = 1.1

over the interval [0, 2π].

Solution

A) Algebraic

First, solve for sin x to get

sin x = -0.1875.

We see that we are looking for all x in [0, 2π] for which sin x = -0.1875. Note that

sin⁻¹(-0.1875) = -0.1886

which is not a solution since it is not in the interval [0, 2π]. We think geometrically about the solutions:

i) solutions are angles with radian measure;

ii) terminal sides of these angles lie in Quadrants I - IV;

iii) the sine of the angles must be negative;

and

iv) their reference angles must be 0.1886 radians.

We know that sin x < 0 in Quadrants III and IV, so given the reference angle of 0.1886, the solutions must be

$$x = \pi + 0.1886 = 3.3302$$

or

$$x = 2\pi - 0.1886 = 6.0946.$$

B) Graphical

On your calculator, enter $y_1 = 3.2\sin x + 1.7$ and $y_2 = 1.1$. Set your x interval from 0 to 2π and your y interval from -6 to 6 (why?); then graph. There are two points of intersection which can be found to be

$$x = 3.3302 \text{ or } x = 6.0946.$$

This method is certainly much easier and we will use the graphical method most of the time. (See Figure 7.11)

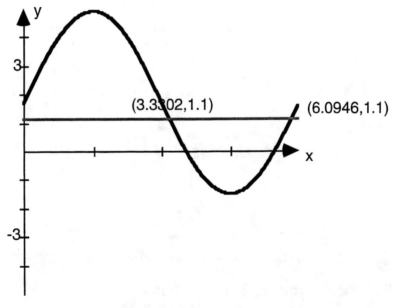

Figure 7.11

* * * *

Example 7.8

Find all solutions to the equation

$$4.1 - 2.3\sin(2x - \pi) = 2.5$$

over the interval $[0, 2\pi]$.

Solution

We solve this equation graphically only. Although it is possible to solve algebraically, there are four solutions and it is much easier to find them all by looking at the graph.

Enter $y_1 = 4.1 - 2.3\sin(2x - \pi)$ and $y_2 = 2.5$, set the x interval from 0 to 2π and the y interval from -7 to 7 (why?), then graph. Use your calculator to find the x coordinates of the four points of intersection to be

x = 1.956, 2.757, 5.097, 5.898. (See Figure 7.12.)

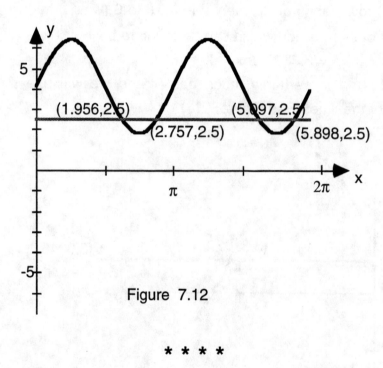

Figure 7.12

* * * *

Example 7.9

Find all solutions to the equation $x = \cos x$.

Solution

We use the graphical method and graph the two functions

$y_1 = x$ and $y_2 = \cos x$.

Figure 7.13 shows the graph with points of intersection labeled. The solution is

x = 0.74.

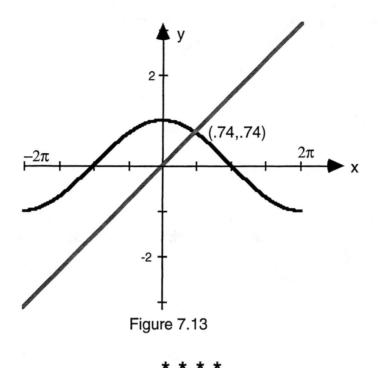

Figure 7.13

* * * *

Class Work

1. Using the algebraic method, find all solutions of the equation

$$3.4\cos x + 1.4 = 2.8$$

in the interval $[0, 2\pi]$.

2. Using the graphical method, find all solutions of the equation

$$\sin x + 2\cos x = 1.42$$

over the interval $[0, 2\pi]$.

* * * *

Exercises

Use reference angles to find all solutions to each equation over the interval $[0, 2\pi]$. Find all solutions to the odd numbered equations.

1. $\sin x = 0.7654$
2. $\cos x = -0.3432$
3. $\tan x = 10.58$
4. $\tan x = -7.71$
5. $2\sin x = 1.8641$

Use the algebraic method to find all solutions to each equation over the interval $[0, 2\pi]$. Find all solutions to the odd numbered equations.

1. $6 + 3\cos x = 4$
2. $\sin x \cos x - 0.78\cos x = 0$
3. $2\sin^2 x - \sin x - 1 = 0$
4. $1.5 - 2.1\tan x = 15$
5. $\tan^2 x - 1 = 0$

Use the graphical method to find all solutions to each equation over the interval $[0, 2\pi]$. Find all solutions to the odd numbered equations.

1. $\sin x = -.34$
2. $1.33 + \cos(2x) = .76$
3. $2\cos x = -1.2$
4. $\tan x = 14$
5. $3.34 - 1.78\sin x = 2.49$
6. $2\cos x - \sin x = 0$
7. $\cos(2x) = 1 - x$
8. $\tan x + x^3 = 0$

* * * *

7.6   Trigonometric Inequalities

A trigonometric inequality is an inequality which involves one or more trigonometric functions. Solving inequalities is easy if you use the graphical technique illustrated in the preceding section: graph each side of the inequality on the calculator and observe that the solution set consists of all x for which one graph is above or below the other, depending on which way the inequality goes. We illustrate these methods by referring to some of the examples from earlier sections.

Example 7.10 (Refer to Example 7.1 in Section 7.5.)
Solve the inequality
$$\sin x \leq \frac{\sqrt{3}}{2}$$
for all x over the interval $[0, 2\pi]$.

Solution

Refer to the graph in Figure 7.9. We are looking for all x for which the graph of y = sin x is on or below the graph of $y = \dfrac{\sqrt{3}}{2}$. We have already found the x-coordinates of the points of intersection of the two graphs to be π/3 and 2π/3, so from this information we see that the solution set is

{x | 0 ≤ x ≤ π/3 or 2π/3 ≤ x ≤ 2π}.

* * * *

Example 7.11 (Refer to Example 7.3 in Section 7.5)

Find all x which satisfy the inequality

3.2sin x + 1.7 > 1.1

over the interval [0, 2π].

Solution

Refer to the graph in Figure 7.11. Here we are looking for all x for which the graph of y = 3.2sin x + 1.7 is above the graph of y = 1.1. We have already found the x-coordinates of the points of intersection of the two graphs to be 3.3302 and 6.0946, so we see that the solution set is {x | 3.3302 < x < 6.0941}.

* * * *

Class Work

Solve each inequality for all x over the interval [0, 2π] and sketch graphs indicating solutions.

1. sin x cos x < 2cos x
2. 4 - sin x ≥ 2

* * * *

Exercises

Solve each inequality for all x over the interval [0, 2π] and sketch graphs indicating solutions.

1. cos x < .67
2. sin x ≥ .35
3. 5 + 2cos x ≤ -2

4. 2sin x cos x > x

5. sin x + cos x ≤ 0.5x

* * * *

7.7   Precipitation

The Nizhoni watershed is quite dry; at the wettest time of year is the end of March (t = 3) when the precipitation rate is 2.5 inches per month, and the end of September (t = 9) when it is 3.0 inches per month. The driest times are late June (t = 6) and the end of December (t = 0 or t = 12). Although in fact it is a bit drier in June than December, to simplify this model we will assume that the rate is the same at both times, 1.65 inches per month. Figure 7.14 below shows a possible precipitation graph.

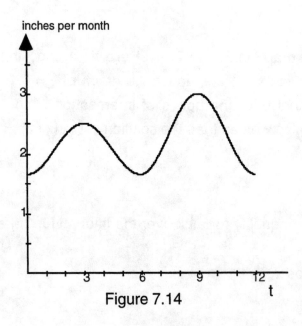

Figure 7.14

In the group work which follows you will derive a function R(t) which describes the rate of precipitation (in inches of water per month) at time t. The sketch above might suggest two functions, one for the first six month and another for the next six month. We will derive two trigonometric functions which together will make up a piece-wise function which will give the rate of precipitation at any time. The steps below will lead you through this procedure.

Group Work

We will determine the equation for each of the two pieces, corresponding to the intervals $0 \le t \le 6$ and $6 \le t \le 12$. Since these will agree for $t = 6$, they can be pieced together to give a function defined for $0 \le t \le 12$.

1. Sketch a graph of the first piece, $0 \le t \le 6$. Recall that the first period has maximum precipitation rate of 2.5 inches per month when $t = 3$ and minimum rate of 1.65 inches per month when $t = 0$ or $t = 6$.

    (a)   Indicate the maximum and minimum values.

    (b)   Determine the period, amplitude, vertical shift and phase shift.

    (c)   Determine an equation which has the properties as shown on the graph. Use the form $y = a + b\cos(ct + d)$.

2. Repeat the process for the second piece, $6 \le t \le 12$. For this interval the maximum precipitation rate is 3.0 inches per month when $t = 9$, with minimum rate of 1.65 inches per month when $t = 6$ or $t = 12$. If the phase shift is more than one period, you might be able to simplify the function. Do so if you can.

3. Write the function R(t) in the form

$$R(t) = \begin{cases} \text{first expression} & \text{if } 0 \le t \le 6 \\ \text{second expression} & \text{if } 6 < t \le 12 \end{cases}$$

and graph the function for one calendar year. Use this function (and its graph) to answer the remaining questions.

4. Predict the precipitation rate on December 15.

5. When will the rate be greater than 2 inches per month?

6. We can estimate the average precipitation for any particular month or portion of a month by determining the rate at the mid-point of the time period. For example, the inches of precipitation for the month of May can be approximated by

    [R(4.5) inches/month] x [1 month];

the number of inches of precipitation for the first half of June is approximated by

[R (5.25) inches/month] x [.5 month].

Estimate the precipitation for:

    (a)   January;

    (b)   May;

    (c)   the last half of October;

    (d)   the period from $t = 2.4$ to $t = 2.7$.

7.   Use R(t) to approximate the total monthly precipitation in inches of water for each month, and sum these figures to determine the total annual precipitation. It will be useful for future work to make a table with two columns, the first indicating the month, the second showing the t value for the month, and the third showing precipitation in that month. (Round your answer to two decimal places.) Use the techniques described in #6 in the previous section.

* * * *

PRECIPITATION SUMMARY

Fill in the required information and save it on a copy of this page; you will need this later. When writing in your answers be sure to include the correct units; this is important and you may forget them later on when you need them.

PRECIPITATION FUNCTION **R**(t) (See question #3.)

MONTHLY AND ANNUAL PRECIPITATION (See question #7.)

MONTH	t	PRECIPITATION (inches)
Jan.		
Feb.		
Mar.		
Apr.		
May		
Jun.		
Jul.		
Aug.		
Sept.		
Oct.		
Nov.		
Dec.		

TOTAL
ANNUAL PRECIPITATION: _____

7.7.1   Rain and Snow

We can now determined at any time t how cold it will be and the rate of precipitation. In this section we will determine how much of this precipitation will fall as rain and how much as snow. This will be important in determining the streamflow since the rain enters the stream almost immediately but the snow doesn't enter the river until the spring snow melt.

1.   Some of the precipitation that falls comes down as rain and some as snow. Use the function R(t), the answer to #8 in the temperature section, and the techniques in #3 above to determine the annual number inches of water from snow. (Round your answer to two decimal places.) Again, a table will be helpful, this time with four columns, the first indicating the interval of time, the second the midpoint, the third the length of the interval, and the fourth the amount of precipitation which is snow.

* * * *

RAIN AND SNOW SUMMARY

Fill in the required information and save it on a copy of this page; you will need this later. When writing in your answers be sure to include the correct units; this is important and you may forget them later on when you need them.

MONTHLY AND ANNUAL SNOW (See question #12.)

MONTH	t	X PORTION OF MONTH	SNOW (inches)
Jan.			
Feb.			
Mar.			
Apr.			
May			
Jun.			
Jul.			
Aug.			
Sept.			
Oct.			
Nov.			
Dec.			

TOTAL ANNUAL SNOW:_____

7.8   Snow   Melt

The rate at which snow is melting, measured in inches per day, depends on temperature as well as other factors. The rate can be approximated by a linear function of surrounding air temperature,

$$M_0 = C(T - 32^\circ),$$

where M_0 denotes the number of inches of water from snow melt per day at the surrounding average air temperature T degrees Fahrenheit, and C is called the melt factor (Hydrology, by Wisler and Brater, John Wiley, N.Y., 1959). Generally the melt factor can vary from .02 to .13; in order to accurately reflect what really happens in the La Plata watershed we choose C = .03, so

$$M_0 = .03\,(T - 32\,).$$

In order to determine streamflow, it will be necessary to know the rate of snow melt over the entire watershed in inches of water per month at time t. Follow the outline below to obtain this function.

Group Work

1.   The function M_0 is given above in inches of water per day. Convert this to a new function M_1 which describes the rate of snow melt in inches of water per month at surrounding air temperature T° F. Round coefficients to four places; use $\frac{365}{12}$ days per month and follow the techniques of #1 in the previous section.

2.   You now have enough information to determine the rate of snow melt at time t. Recall the function A(t) for temperature derived in the preceding section. Form a composite function by substituting this for T in the function M_1 from #1 above; then simplify. The result should be a function M(t) which describes the rate of snow melt in inches of water per month at time t. (Round coefficients to two places.)

The next two parts enable you to determine the domain of the function M(t).

3.   Note that $M_0 = .03(T - 32)$ only makes sense if $T \geq 32^\circ$, so recall your answer to #8 from the **Temperature** section and use this to specify the left-hand end point of the domain of M(t). (Round values of t to one decimal place.)

4.   Recall your answer to #1 from the **Rain and Snow** section and use this and the function M(t) to determine when all of the snow in the watershed will be melted. First approximate snow melt for each appropriate month or portion of a month; use the mid-point of the interval to estimate the rate of snow melt for that interval and round answers to one place. Your answer is obtained when the cumulative number of inches of water from melting snow approximately equals the total number of inches of water from snow determined in #4 in the previous section. This will determine the right-hand end point of the domain of M(t). See the Table on the Summary page.

5.   What is the interval which is the domain of M(t)?

6.   Graph the function M(t) and indicate its domain.

* * * *

SNOW MELT SUMMARY

Fill in the required information and save it on a copy of this page; you will need this later. When writing in your answers be sure to include the correct units; this is important and you may forget them later on when you need them.

SNOW MELT FUNCTION M(t) (See question #2.)

MONTHLY AND CUMULATIVE SNOW MELT (inches) (See question #4.)

MONTH	t	M(t)	X PORTION OF MONTH	MONTH'S SNOW MELT	CUMULATIVE SNOW MELT

DOMAIN OF M(t), i. e., the interval of time beginning when snow starts to melt and ending when all the snow is melted (See question #5.)

* * * *

CHAPTER EIGHT
SNOW FED RIVERS: STREAMFLOW

Introduction

Up until this point, we have been talking about depth of rain and snow. However, the water which feeds the Nizhoni River drains from its entire watershed so we need to know the volume of water over this region to determine the streamflow. We will modify both the precipitation function R(t) (which describes precipitation rate in terms of inches per month) and the sow melt function (which describes rate of snow melt in inches per month) to get related functions W(t) and S(t) which describe the rates of precipitation and snow melt over the whole watershed in million cubic feet of water per month.

8.1   Area of the Watershed

In order to accomplish our goal, we first need to determine the area of the watershed. The watershed can be approximated by an irregular shaped figure for which there is no convenient formula for determining the area. A useful trigonometric formula which enables us to find this area is known as the *Law of Cosines* : in the triangle shown (Figure 8.1),

$$a^2 = b^2 + c^2 - 2bc\cos\alpha.$$

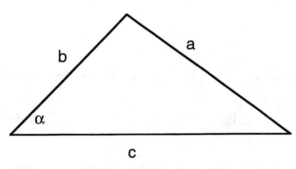

Figure 8.1

The following example illustrates how this can be used to determine the area of a triangle when its sides are known.

Example 8.1

Find the area of the triangle shown (Figure 8.2).

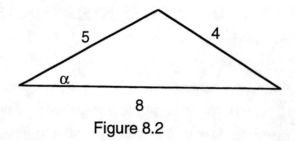

Figure 8.2

Solution

First. we need to know the height. Drop a perpendicular line from the top vertex to the base; the length of this line is the height. Using the Law of Cosines, we can find the angle α and then use this to determine h. (See Figure 8.3 below.)

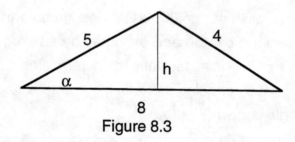

Figure 8.3

$$4^2 = 5^2 + 8^2 - 2 \cdot 5 \cdot 8 \cos \alpha$$
$$80 \cos \alpha = 25 + 64 - 16$$
$$\cos \alpha = \frac{73}{80}$$
$$\alpha = \cos^{-1} \frac{73}{80}$$
$$h = 5 \sin \alpha = 2.05 \text{ (rounded)}$$

Now we get the area of the triangle (one-half base x height) to be

$$A = \frac{1}{2} (2.05)(8) = 8.20 .$$

* * * *

Group Work

The shape of the watershed can be approximated by Figure 8.4 as shown. (Dimensions are given in miles.) Use the triangle-solving techniques illustrated above to determine the area of the watershed. Round all intermediate calculations to four decimal places, and round the final answer to the nearest square mile.

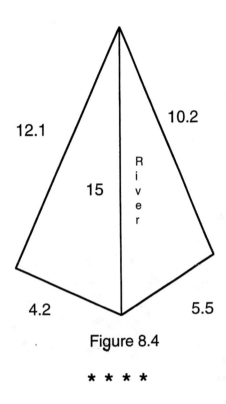

Figure 8.4

* * * *

WATERSHED SUMMARY

Fill in the required information and save it on a copy of this page; you will need this later. When writing in your answers be sure to include the correct units; this is important and you may forget them later on when you need them.

AREA OF THE WATERSHED_____

8.2   Solving Triangles

In this chapter we used the law of cosines to determine the area of a triangular region when the lengths of the sides were known. The law of cosines provided a method for determining the angles of a triangle and from this information we could determine the altitude and then the area. In this section we will take a closer look at the *law of cosines* and also the *law of sines*.

If enough information is given about the sides and angles of a triangle, the remaining parts can be determined. The law of cosines is useful when all the sides of a triangle are known or when two sides and the angle between them are known. On the other hand the law of sines can be used when either

i)   two angles and the included side are given, or

ii) two sides and another angle, not the one between them, are known.

The second case, as you will see, does not always lead to one solution.

8.2.1   The Law of Cosines

The Law of Cosines. In the triangle ABC shown (Figure 8.5),
$$a^2 = b^2 + c^2 - 2bc\cos\alpha,$$
$$b^2 = a^2 + c^2 - 2ac\cos\beta,$$
$$c^2 = a^2 + b^2 - 2ab\cos\gamma.$$

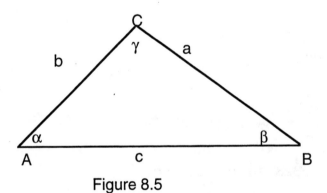

Figure 8.5

We show the derivation of this formula in Appendix A; here we illustrate its use in two examples.

Example 8.2

A triangle has sides of length 5, 8 and 11. Determine the angles.

Solution

Label $a = 5$, $b = 8$ and $c = 11$. We can use the different forms of the law of cosines to determine the angles α, β, and γ. These angles can be measured in degrees or radians; for this example we chose radian measure.

$$5^2 = 8^2 + 11^2 - 2 \cdot 8 \cdot 11 \cos \alpha$$
$$176 \cos \alpha = 64 + 121 - 25$$
$$\cos \alpha = \frac{160}{176}$$
$$\alpha = \cos^{-1} \frac{160}{176}$$
$$\alpha = .4297 \text{ radians (rounded).}$$

In a similar fashion we can determine β and γ:
$$8^2 = 5^2 + 11^2 - 2 \cdot 5 \cdot 11 \cos \beta$$
$$110 \cos \beta = 25 + 121 - 64$$
$$\cos \beta = \frac{82}{110}$$
$$\beta = \cos^{-1} \frac{82}{110}$$
$$\beta = .7296 \text{ radians (rounded);}$$

and
$$11^2 = 5^2 + 8^2 - 2 \cdot 5 \cdot 8 \cos \gamma$$
$$80 \cos \gamma = 25 + 64 - 121$$
$$\cos \gamma = \frac{-32}{80}$$
$$\gamma = \cos^{-1} \frac{-32}{80}$$
$$\gamma = 1.9823 \text{ radians (rounded).}$$

Notice that the sum of the angles is
$$\alpha + \beta + \gamma = .4297 + .7296 + 1.9823 = 3.1416 \text{ radians.}$$
Recall that π radians = 180º, so this is just what you expect.

* * * *

Example 8.3

A triangle has sides a = 12, b = 10 and angle $\gamma = 113°$. Determine the remaining side.

Solution

Make sure you change your calculator mode from radians to degrees. We use the form $c^2 = a^2 + b^2 - 2ab\cos\gamma$,

$$c^2 = 12^2 + 10^2 - 2 \cdot 12 \cdot 10\cos 113$$

$$c^2 = 18.3787$$

$$c = 18.38 \text{ (rounded to 2 places)}.$$

* * * *

Class Work

1. A triangle has sides b = 17.1, c = 11.5 and angle $\alpha = 37.2°$. Determine the remaining side. Round your answers to one decimal place.

2. A triangle has sides a = 94, b = 46 and c = 52. Determine the angles. Round your answers to two decimal places.

* * * *

8.2.2   The Law of Sines

When either

i)   two angles and a side are given, or

ii) two sides and another angle, not the one between them, are known, the law of cosines cannot be used but the law of sines comes to the rescue.

Law of Sines: In the triangle ABC (Figure 8.6),

$$\frac{\sin\alpha}{a} = \frac{\sin\beta}{b} = \frac{\sin\gamma}{c}.$$

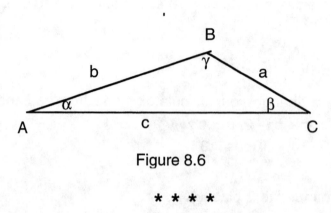

Figure 8.6

* * * *

Before illustrating the use of the law of sines, we look at several different situations that can occur. If two angles and any side are given then the third angle can be found and there is a unique triangle as shown in Figure 8.7.

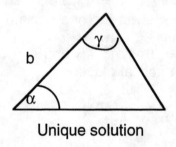

Unique solution

Figure 8.7

If two sides and another angle, not the one between them, are known then several possibilities can occur as shown in Figures 8.8, 8.9, 8.10, and 8.11. The angle and sides which are labeled are the ones that are given.)

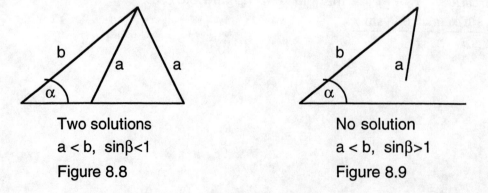

Two solutions   No solution

$a < b$, $\sin\beta < 1$ $a < b$, $\sin\beta > 1$

Figure 8.8   Figure 8.9

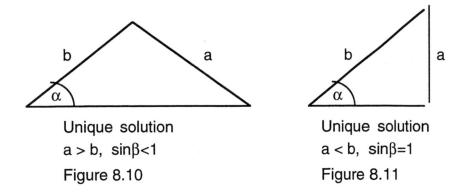

Unique solution   Unique solution
$a > b$, $\sin\beta < 1$ $a < b$, $\sin\beta = 1$
Figure 8.10   Figure 8.11

We show the derivation of the law of sines in Appendix A; here we illustrate its use by giving some examples.

Example 8.4
Given $\alpha = 22^\circ$, $\beta = 49^\circ$, and $c = 14.3$ determine the remaining parts.

Solution
It easy to determine the third angle, $\gamma = 180 - (22 + 49) = 109^\circ$. Now we use the law of sines to determine the remaining sides. In order to find a we use the equation
$$\frac{\sin\alpha}{a} = \frac{\sin\gamma}{c}.$$
Substitute for α, γ and c,
$$\frac{\sin 22^o}{a} = \frac{\sin 109^o}{14.3}$$
and solve for $a = 5.67$ (rounded to two places).

To find b we use the equation
$$\frac{\sin\beta}{b} = \frac{\sin\gamma}{c}.$$
Substitute for β, γ and c,
$$\frac{\sin 49^o}{b} = \frac{\sin 109^o}{14.3}$$
and solve for $b = 11.41$ (again rounded to two places).

$$\ast \; \ast \; \ast \; \ast$$

Example 8.5
Given $\alpha = 31^\circ$, $a = 7.8$, $c = 12.5$ find all remaining parts of the triangle.

Solution

This problem is different because there are two possible triangles which have the given angle and sides. (See Figure 8.12.)

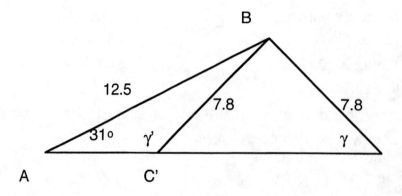

Figure 8.12

To determine γ we use the equation

$$\frac{\sin \alpha}{a} = \frac{\sin \gamma}{c}$$

since α, a and c are known. Substitute these values to get

$$\frac{\sin 31^{\circ}}{7.8} = \frac{\sin \gamma}{12.5},$$

then solve for $\sin\gamma$,

$\sin\gamma = .8254$ (rounded to 4 places).

One solution to this equation is

$\gamma = 55.6^{\circ}$.

Referring to Figure 6.27 (above) we see that the other solution is

$\gamma' = 180^{\circ} - \gamma = 124.4^{\circ}$.

We continue to solve the big triangle (Figure 8.13).

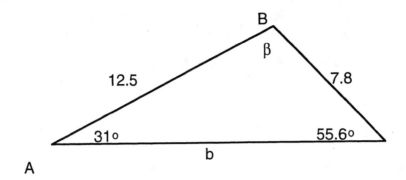

Figure 8.13

The angle $\beta = 180 - (31+55.6) = 93.4^0.$ Use the equation
$$\frac{\sin 31^o}{7.8} = \frac{\sin 93.4^o}{b}$$
to find b = 15.1.

We solve the smaller triangle (Figure 8.14) in a similar manner to get $\beta'= 24.6^0$ and b' = 6.3.

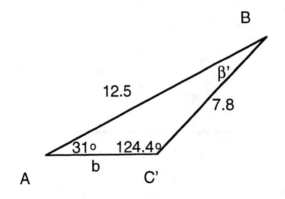

Figure 8.14

* * * *

Example 8.6

If $\alpha = 44^o$, a = 12.3, c = 20.5 find all the remaining parts.

Solution

First solve the equation
$$\frac{\sin 44^o}{12.3} = \frac{\sin \gamma}{20.5}$$

for

$$\sin\gamma = 1.1578.$$

Since this has no solution, there is no triangle with the given parts.

* * * *

Class Work

1. A triangle has angles $\beta = 14.5^0$, $\gamma = 83.2^0$, and side $c = 18.9$. Determine b.

2. A triangle has sides $a = 3.5$ and $b = 4.2$, and angle $\beta = 75^0$. Determine all remaining parts of the triangle.

* * * *

Exercises

For each of the problems below, determine all missing parts of the triangle. Refer to the triangle shown.

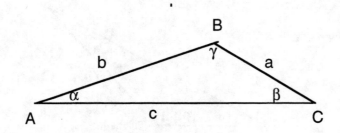

1. $\alpha = 33^0$, b = 52.1, c = 73.4
2. $\beta = 57^0$ $\gamma = 103^0$, c = 98.6
3.   a = 14.9, b = 27, $\beta = 12^0$
4.   a = 12, b = 29, c = 39
5. $\alpha = 21^0$, a = 25, b = 41
6. $\beta = 48.1^0$, b = 21.3, a = 71.4
7. $\alpha = 30^0$, b = 11.2, a = 15
8. $\gamma = 33^0$, c = 45.45, a = 29
9.   Write the proof of the Law of Cosines when written in the form
      $b^2 = a^2 + c^2 - 2ac\cos\beta$.
10.   Write the proof of the Law of Sines for the case

$$\frac{\sin \alpha}{a} = \frac{\sin \gamma}{c}.$$

* * * *

8.3   Rain And Snow Over The Entire Watershed

Now that you have determined the area of the watershed it is possible to determine the precipitation rate over the entire watershed as cubic feet per month. Before you do this, it will help to think for a minute about volume. Imagine that one inch of rain that falls uniformly over one square mile. The volume of rain is the volume of a box of water with base 1 square mile and height one inch. The volume of that box in cubic feet is

$$1 \text{ square mile x 1 inch} = 5280^2 \text{ square feet x 1 inch}$$

$$= 27.8784 \times 10^6 \text{ square feet x } \frac{1}{12} \text{ foot}$$

$$= \frac{27.8784}{12} \times 10^6 \text{ cubic feet}$$

$$= 2.3232 \times 10^6 \text{ ft}^3.$$

Thus one inch of rain produces $2.3232 \times 10^6 \text{ ft}^3$ of water over each square mile of the watershed.

Group Work

Follow the steps below to determine a function W(t) which describes the precipitation rate at time t in terms of million cubic feet per month.

1.   Determine the number of cubic feet ($\times 10^6$) of water one inch of precipitation will produce over the whole watershed.

2.   Recall the annual precipitation (in inches) from your Precipitation Summary. Use this and the factor determined in #1 to determine the total volume of precipitation over the watershed each year.

3.   Use the factor determined in #1 to convert the function R(t), which measures inches of rainfall per month, to a new function W(t) which measures cubic feet of water per month over the entire watershed at time t. Leave your function in product form.

4. Convert the function M(t), which measures inches of water from snow melt per month, to a new function S(t) which measures cubic feet of water per month over the entire watershed at time t. Leave your function in product form.

5. Graph the functions W(t) and S(t).

* * * *

RAIN AND SNOW OVER THE ENTIRE WATERSHED SUMMARY

Fill in the required information and save it on a copy of this page; you will need this later. When writing in your answers be sure to include the correct units; this is important and you may forget them later on when you need them.

VOLUME OF ANNUAL PRECIPITATION (See question #2)

CONVERTED PRECIPITATION AND SNOW MELT FUNCTIONS (See question #4)

8.4   Streamflow

Streamflow is defined as the volume of water which passes a cross-section of a stream in a specified unit of time. Units are usually cubic feet per second (cfs) but could be otherwise, such as cubic inches per minute or cubic meters per day, etc. Streamflow will be different at various points on a stream and at different times of the year, so point of measurement as well as time of year must be designated. For example, to say that the streamflow of the La Plata River at Hesperus, Colorado, on April 15 is 83 cfs means that on that day at Hesperus, 83 cubic feet of water passed down the La Plata in one second. It is important to be able to predict flow for periods during the year in order to monitor water storage or to prepare for floods or droughts and other things of this nature. In this section of the study, we will predict streamflow for Nizhoni River. Our prediction will be based on information about precipitation, temperature and snow melt derived previously.

We consider three ways in which precipitation is dispersed, *runoff*, *ground water*, and *evaporation*. Runoff refers to water which runs over the surface of the Ground directly into the stream; ground water is water which seeps into the ground; and evaporation refers to water which, well..., evaporates. So runoff is a direct and immediate contribution to streamflow; ground water eventually seeps into the stream thus being an indirect contributor to streamflow. Evaporation, of course, contributes nothing. We make the following assumptions regarding precipitation and the water from melting snow:

   * 80% is runoff;

   * 15% is ground water;

   * 5% is evaporated.

For any stream, *base flow* is water obtained from ground water, and since ground water seeps slowly into the stream, we consider base flow for our river to be constant year round, i. e., the same each month. Therefore, base flow is sort of a minimum flow for the stream, rain and snow melt are like "extra" water.

You now have all the information needed to write a function which describes streamflow. This will be a piecewise function; follow the steps below to obtain its definition.

Group Work

1.   Determine the periods of the year when base flow is the only water in Nizhoni River. (See Figure 8.14.)

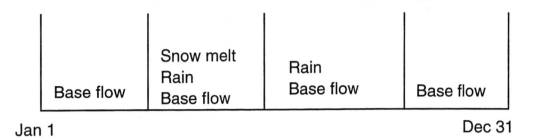

Figure 8.14

2.   Determine each of the following (see Figure 8.14):

      a. the period of the year during which, in addition to base flow, the stream receives water from both rain and snow melt;

      b. the period of the year during which the only source of water other than base flow received by the stream is from rain.

3.   We consider the base flow to be constant each month. Base flow comes from ground water, so the total annual base flow is 15% of the annual volume of precipitation, and we assume the monthly base flow will be one-twelfth of this. Determine monthly base flow for the Nizhoni River. (Write your answer x10^6 cubic feet and round to two decimal places.)

4.   Now put the pieces together. Use the assumptions made in this section to write the piecewise function F(t) which describes streamflow for Nizhoni River in cubic feet (x10^6) in month t, $0 \le t \le 12$. When simplifying, round coefficients to two decimal places. (See Figure 6.32.)

5.   Graph the piecewise function F(t) and indicate the interval over which each piece is defined.

6.   Use the function F(t) to approximate monthly flows for the river, i. e., the total amount of water which flows through each month. This must be done separately for each piece of the function F(t): make a table with four columns, the first indicating the interval of time, the second the mid-point, the third the length of the interval and the fourth indicating the streamflow for the month or the portion of the month. Now make a table which indicates total streamflow for

each month of the year (see summary sheet). (Round answers to the nearest integer.)

* * * *

STREAMFLOW SUMMARY

Fill in the required information and save it on a copy of this page; you will need this later. When writing in your answers be sure to include the correct units; this is important and you may forget them later on when you need them.

STREAMFLOW FUNCTION **F**(t) (See question #4.)

MONTHLY AND ANNUAL STREAMFLOW (x10^6) CUBIC FEET

MONTH	t	STREAMFLOW
Jan.		
Feb.		
Mar.		
Apr.		
May		
Jun.		
Jul.		
Aug.		
Sep.		
Oct.		
Nov.		
Dec.		

TOTAL ANNUAL STREAMFLOW_____

8.5 Identities

An identity is an equation which is true for all values of the variable for which the functions are defined. The identities in this section are ones that you will encounter in your study of calculus. It is important that you remember these and be able to use them to simplify trigonometric expressions; examples and exercises are provided.

8.5.1 Basic Identities

We return to the circle with center at the origin and radius r > 0 and recall the definitions of the sine and cosine; if t is the radian measure of an angle in standard position and (x, y) the point of intersection of the terminal side of the angle with the circle then

x = rcos t and y = rsin t.

When r = 1 the point (x,y) = (cost,sint), and since the equation of the unit circle is

$$x^2 + y^2 = 1,$$

we have the basic identity

(1) $\cos^2 t + \sin^2 t = 1.$

(We point out that $\cos^2 t$ means $(\cos t)^2$ and $\sin^2 t$ means $(\sin t)^2$; it is standard to abbreviate powers of trigonometric functions in this way.)

Also, other basic identities follow immediately from the definitions of the functions.

(2) $sin t = \dfrac{1}{csc t}$

(3) $cos t = \dfrac{1}{sec t}$

(4) $tan t = \dfrac{1}{cot t}$

(5) $tan t = \dfrac{sin t}{cos t}$

(6) $cot t = \dfrac{cos t}{sin t}$

(Of course, these identities are valid wherever denominators are non-zero.)

We can use these identities to derive new identities, as illustrated below. We begin with the basic identity

$$\cos^2 t + \sin^2 t = 1;$$

divide both sides of the equation by $\cos^2 t$ and simplify,

$$\frac{\cos^2 t + \sin^2 t}{\cos^2 t} = \frac{1}{\cos^2 t}$$

$$\frac{\cos^2 t}{\cos^2 t} + \frac{\sin^2 t}{\cos^2 t} = \frac{1}{\cos^2 t}$$

$$1 + \tan^2 t = \sec^2 t$$

From this we get a fourth basic identity

(7) $1 + \tan^2 t = \sec^2 t$.

Other identities of interest are:

(8)   sin(-t)= -sint;

(9)   cos(-t) = cost;

(10)   tan(-t) = -tant;

(11) $\cos(\frac{\pi}{2} - t) = \sin t$;

(12) $\sin(\frac{\pi}{2} - t) = \cos t$.

$$* \quad * \quad * \quad *$$

We illustrate why basic identities #8 and 9 are true when α is an acute angle. In Figure 8.15 the angles α and $-\alpha$ are both shown in standard position.

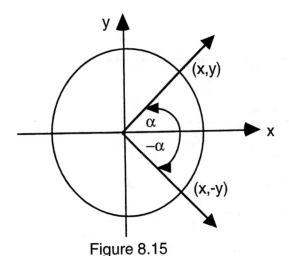

Figure 8.15

It should be clear that if (x, y) is the point of intersection of the terminal side of α with the unit circle, then the point (x, -y) will be the point of intersection of the terminal side of the angle -α and the unit circle. Thus,

$$\cos\alpha = x = \cos(-\alpha),$$

and

$$\sin\alpha = y, \sin(-\alpha) = -y,$$

or

$$-\sin\alpha = \sin(-\alpha).$$

* * * *

Below are some examples of simplification of trigonometric expressions using the basic identities.

Example 8.7

Simplify the expression
$$\frac{sin\theta}{1-cos^2\theta}$$
to one which is a single term, is not a fraction, and involves only one trigonometric function.

Solution

Use the identity
$$\cos^2 t + \sin^2 t = 1$$
in the form
$$\sin^2 t = 1 - \cos^2 t$$
to replace the denominator. We now have
$$\frac{sin\theta}{sin^2\theta}.$$
After canceling out the sines, we get
$$\frac{1}{sin\theta},$$
which, by identity (2) is
$$csc\theta.$$

* * * *

Example 8.8

Simplify the expression

$$\sec\theta - \cos\theta$$

to one which is a single term, is not a fraction, and is a product of two trigonometric functions.

Solution

Use identity (3) to replace $\sec\theta$ to get

$$\frac{1}{\cos\theta} - \cos\theta,$$

then add the two terms together,

$$\frac{1 - \cos^2\theta}{\cos\theta}.$$

Rewrite this expression and then use identity (5) to get

$$\frac{\sin\theta}{\cos\theta}\sin\theta = \tan\theta\sin\theta.$$

* * * *

Example 8.9

Simplify the expression

$$\frac{\sec t - \tan t}{\cos t} - \frac{\cos t}{\sec t + \tan t}$$

to one which is a single fraction, and involves only one trigonometric function.

Solution

First, find the common denominator and add the two fractions together,

$$\frac{(\sec t - \tan t)(\sec t + \tan t)}{(\sec t + \tan t)\cos t} - \frac{\cos^2 t}{(\sec t + \tan t)\cos t} = \frac{\sec^2 t - \tan^2 t}{(\sec t + \tan t)\cos t} - \frac{\cos^2 t}{(\sec t + \tan t)\cos t}$$

$$= \frac{\left(\sec^2 t - \tan^2 t\right) - \cos^2 t}{(\sec t + \tan t)\cos t}$$

and use identity (7) to replace $\sec^2 t - \tan^2 t$. We have

$$\frac{1 - \cos^2 t}{(\sec t + \tan t)\cos t} = \frac{\sin^2 t}{(\sec t + \tan t)\cos t}.$$

Now consider the factor $(\sec t + \tan t)$ in the denominator. By using identities (3) and (5), this becomes

$$\frac{1}{cos\,t}+\frac{sin\,t}{cos\,t}=\frac{1-sin\,t}{cos\,t},$$

so that we have

$$\frac{sin^2\,t}{(sec\,t+tan\,t)cos\,t}=\frac{sin^2\,t}{\left(\dfrac{1-sin\,t}{cos\,t}\right)cos\,t}=\frac{sin^2\,t}{1-sin\,t},$$

which is in the desired form.

* * * *

Exercises

1. Simplify the expression
$$\frac{tan\,t}{1+tan^2\,t}$$
to one which is not a fraction and is a product of two trigonometric functions.

2. Simplify the expression
$$sec\,t\,tan\,t+sec^2\,t$$
to one which is a single fraction and involves only one trigonometric function.

3. Simplify the expression
$$\left(\frac{1}{sin\,t}-sin\,t\right)sin\,t$$
to one which is not a fraction and involves only the cosine function.

4. Simplify the expression
$$(sec\,x-cos\,x)cos\,x$$
to one which is not a fraction and involves only the sine function.

5. Simplify the expression
$$tan^2\,t-sin^2\,t$$
to one which is a product of tangent and sine.

6. Simplify the expression
$$\frac{1-cos^2\,x}{sin^2\,cos\,x}$$
to one which is not a fraction and involves only one trigonometric function.

7. Use identities #8 and 9 to derive #10 for tan(-t).

8. For $0 < t < \pi/2$, illustrate why identity #11 is true.

9. For $0 < t < \pi/2$, illustrate why identity #12 is true.

* * * *

8.5.2   Sum and Difference Formulas.

These formulas are identities which we list here and prove in Appendix A.

(1) $\cos(\alpha - \beta) = \cos\alpha\cos\beta + \sin\alpha\sin\beta$.

(2) $\cos(\alpha + \beta) = \cos\alpha\cos\beta - \sin\alpha\sin\beta$.

(3) $\sin(\alpha + \beta) = \sin\alpha\cos\beta + \cos\alpha\sin\beta$.

(4) $\sin(\alpha - \beta) = \sin\alpha\cos\beta - \cos\alpha\sin\beta$.

We state without proof some additional useful identities. The derivations are left as exercises.

(5) $\tan(\alpha + \beta) = \dfrac{\tan\alpha + \tan\beta}{1 - \tan\alpha\tan\beta}$.

(6) $\tan(\alpha - \beta) = \dfrac{\tan\alpha - \tan\beta}{1 + \tan\alpha\tan\beta}$.

(7) $\sin(2\alpha) = 2\sin\alpha\cos\alpha$.

(8) $\cos(2\alpha) = \cos^2\alpha - \sin^2\alpha = 2\cos^2\alpha - 1 = 1 - 2\sin^2\alpha$.

$$\ast \quad \ast \quad \ast \quad \ast$$

The following examples illustrate some of the uses for the sum and difference formulas.

Example   8.10   Prove that $\sin(\alpha + \pi) = -\sin\alpha$.

Solution   Use the formula for the sine of the sum of angles with $\beta = \pi$.

$$\sin(\alpha + \pi) = \sin\alpha\cos\pi - \cos\alpha\sin\pi$$
$$= \sin\alpha \cdot -1 - \cos\alpha \cdot 0$$
$$= -\sin\alpha.$$

$$\ast \quad \ast \quad \ast \quad \ast$$

Example 8.11   Given that α and β are acute angles, $\cos\alpha = .6$, and $\sin\beta = .4$, determine $\sin(\alpha+\beta)$.

Solution: For this we need the formula $\sin(\alpha + \beta) = \sin\alpha\cos\beta + \cos\alpha\sin\beta$. The values are given for two of the quantities on the right-hand side of the equal sign, $\cos\alpha = .6$ and $\sin\beta = .4$. All that is left is to find the two missing pieces on the right. First we find $\sin\alpha$,

$$\cos \alpha = .6$$

$$\sin^2 \alpha = 1 - \cos^2 \alpha = 1 - .36$$

$$\sin^2 \alpha = .64$$

$$\sin \alpha = \pm \sqrt{.64} = \pm .8.$$

But α is acute, so if it is placed in standard position, the terminal side of the angle is in the first quadrant. Therefore sinα is positive and so

$$\sin \alpha = .8.$$

We determine cosβ in a similar fashion:

$$\sin \beta = .4$$

$$\cos^2 \beta = 1 - \sin^2 \beta = 1 - .16$$

$$\cos^2 \beta = .84$$

$$\cos \beta = +\sqrt{.84} = .92 \text{ (rounded to two decimal places)}.$$

It is easy now to determine sin(α+β) by substituting into the formula:

$$\sin(\alpha+\beta) = (.8)(.92) + (.6)(.4) = .976.$$

＊ ＊ ＊ ＊

Example 8.12   Given that α is an obtuse angle, β is an acute angle, sinα = .6, and sinβ = .2, determine cos(α–β).

Solution   When placed in standard position, the terminal side of α is in the second quadrant and the terminal side of β is in the first quadrant. This indicates that cosα must be negative, while cosβ is positive. Therefore

$$\cos \alpha = -\sqrt{1 - \sin^2 \alpha} = -\sqrt{1 - .36} = -\sqrt{.64} = -.8, \text{ and}$$

$$\cos \beta = +\sqrt{1 - \sin^2 \beta} = \sqrt{1 - .04} = \sqrt{.96} = .98 \text{ (rounded)}.$$

Then cos(α–β) = (-.8)(.98) + (.6)(.2) = -.664.

＊ ＊ ＊ ＊

Example 8.13   If cos(2a) = .6, determine cos(α).

Solution: This can be solved using the double angle formula $\cos(2\alpha) = 2\cos^2 \alpha - 1$.

$$.6 = 2\cos^2\alpha - 1$$
$$1.6 = 2\cos^2\alpha$$
$$.8 = \cos^2\alpha$$
$$\cos\alpha = \pm\sqrt{.8} = \pm.89.$$

(Notice the $\pm$ sign; the correct sign cannot be determined unless you are given more information about α.)

$$\ast\ \ast\ \ast\ \ast$$

Example 8.14   (a)   Prove that $\cos\alpha = \pm\sqrt{\dfrac{\cos(2\alpha)+1}{2}}$.

(b)   Use this to prove the half-angle formula, $\cos\left(\dfrac{\theta}{2}\right) = \pm\sqrt{\dfrac{1+\cos\theta}{2}}$.

Solution   (a)   This is an immediate consequence of the double angle formula
$$\cos(2\alpha) = 2\cos^2\alpha - 1.$$
Solve this equation for $\cos\alpha$:
$$\cos(2\alpha) = 2\cos^2\alpha - 1$$
$$2\cos^2\alpha = 1 + \cos(2\alpha)$$
$$\cos^2\alpha = \frac{1+\cos(2\alpha)}{2}$$
$$\cos\alpha = \pm\sqrt{\frac{1+\cos(2\alpha)}{2}}.$$

(b)   This formula, one of the half-angle formulas, follow immediately from part (a), with the substitution $\alpha = \dfrac{\theta}{2}$.

$$\ast\ \ast\ \ast\ \ast$$

Exercises

For exercise 1 - 4, assume that α and β are both acute. Use the given information to find what you are asked.

1. $\sin\alpha = .9$, $\cos\beta = .1$, determine $\sin(\alpha+\beta)$.

2. $\cos\alpha = .4$, determine $\cos(2\alpha)$.

3.   tanα = .3, determine tan(2α).

4.   sinα = .2, sinβ =.4; determine cos(α+β)

For exercises 5 - 8, assume that α is acute and β is obtuse. Use the given information to find what you are asked.

5.   sinα = .1, cosβ = -.2, determine cos(α-β).

6.   sinβ = .4, determine cos(2β).

7.   tan(2α)=1; determine tanα.

8.   cosβ = -.6, sinα = .8, determine sin(β-α).

Verify the following identities.

9.   sin(α+π/2) = cosα.

10.   cos(π+ α)=-cosα

11.   Use the identities for sin(α + β) and cos(α + β) to derive identity #5 for tan(α + β). Hint: after applying these two identities, divide both numerator and denominator by an appropriate term.

12.   Write -β as +(-β) and derive identity #6 for tan(α + β).

13.   Use the identity for sin(α + β) to derive identity #7 for sin(2α).

14.   Derive all three forms in identity #8 for cos(2α).

* * * *

8.6   Designing a Reservoir

In this section we once again return to the village of River City. Recall that the village is growing and with this, of course, there is a proportionate increase in water demand. In order to accommodate this increase, it may be necessary to construct a reservoir, and details of this project are dependent upon both population growth and availability of water. We have completed the determination of streamflow for the Nizhoni River and are now ready to attack the problem of future water needs for River City.

8.6.1   Withdrawal Regulations and Water Needs

Upon first settling, the people of the village daily withdrew the amount of water necessary for their personal use. However, as the population grew the residents who lived further downstream noticed that at certain times during the year less and less water was actually getting to them. This was obviously a cause for concern. They observed that in spring and fall, there was ample water flowing through but that in late fall and early winter, there was not much at all getting downstream. It became clear that the increasing population was drawing more and more water from the Nizhoni River, thus decreasing the water flowing downstream and eventually depriving the downstream settlers of sufficient water in the latter part of the year. To protect the rights of downstream users (other villages and towns, farmers, ranchers, etc.) regulations were put on the maximum amount of water that can be withdrawn from the river. The regulations restrict River City to a maximum withdrawal of 10% of streamflow each month. Sometimes the Nizhoni River provided more than enough water, sometimes barely enough. Hence, it could be desirable to save the excess water which flows through in spring and fall for use in the drier months of the year, i. e., build a reservoir. But how big should it be? If indeed the residents are to assume the seemingly necessary task of constructing a reservoir, in light of the increasing population it would be wise to build it so that it would serve the town for some years to come. The village decides to build a reservoir provided that it is possible to build one large enough to handle the water needs of the community for at least the next thirty years. In order to make the final decision some important questions needed answering.

> * What is the projected population for the next 30-35 years?
> * How many people can the river support if no reservoir is built?
> * How many people can the river support if a reservoir is built?
> * When must the reservoir be completed in order that the residents not be without water?
> * If the reservoir is built to capacity, for how many years will it serve the needs of the community?

These questions are addressed in the steps below. Work through each of these steps (you will need answers obtained from previous work in this study) to reach your conclusions about provision of water for future residents of River City. When you have completed the steps below, write a summary report which answers each of the above questions.

Group Work

We will determine projected water needs based on the population model.

1. Use the population function derived in Chapter 4 to predict the village population in 35 years.

2. Assuming a constant monthly amount of water is used each month, what is the total amount of water used monthly by the residents presently? Can enough water be withdrawn each month from Nizhoni River to provide for this usage?

Next we determine how many people the river can support, with and without a reservoir.

3. Determine how many people the river can support each month if no reservoir is built. Assume that the residents use a constant amount of water each month. When will the population reach this amount?

4. Determine the maximum amount of water that can be withdrawn from the reservoir each month if the regulations are followed. The table shown at the end of this problem set will help you answer this question. The streamflow column repeats the information derived in the streamflow section.

5. How many people can be supported by the river if all the water which can be withdrawn annually is saved for use when needed.?

6. When will the population reach this level?

7. Summarize your work in a report which answers the questions posed above. Should the town build the reservoir? If so, when must it be completed and for approximately how many years will it supply enough water for all of the residents.

MONTH	STREAMFLOW	WITHDRAWAL
1		
2		
3		
4		
5		
6		
7		
8		
9		
10		
11		
12		
TOTAL		

8.6.2   Building a Reservoir

Follow the steps outlined below to determine the capacity of a reservoir which will serve the citizens of River City for as many years as possible. The reservoir should support the population which you determined in #5 in the preceding section.

Group Work

Fill in the table below to determine how big the reservoir should be and how much water it will hold at different times of the year. The first column in this new table comes from the table in the preceding section.

1.   i) Determine the monthly usage for the population determined in #5 in the preceding section. Assume the water usage is constant each month.

ii) Determine what the surplus or deficit is for each month; i.e. what is the difference between the amount of water used and the amount withdrawn from the river each month. Use positive numbers to denote surplus (when more water is withdrawn than used) and negative numbers to denote deficit.

2.   Determine the minimum volume of the reservoir which must be constructed to hold the amount of water necessary to store in order to support the maximum number of people of River City. Since you want to have a little extra room, round up to the nearest 5 million cubic feet.

3.   The last column of the table constructed indicates the volume of water in the reservoir at the end of the indicated month. The volume of the reservoir is slightly more than the amount needed, so assuming the reservoir is filled to capacity there will always be a certain amount of water remaining (this will be the minimum volume of water, or the difference between the actual capacity and the required capacity you determined in #2). Complete the table beginning with the first month with surplus water and assume that the volume of water is minimum at this time.

4.   Summarize your work in the form of a short report. Be sure to designate the volume of the reservoir, the number of people it will support under current usage, and the number of years before the reservoir will become inadequate due to population increase.

MONTH	ALLOWABLE WITHDRAW	MONTHLY USAGE	SURPLUS or DEFICIT	VOLUME OF WATER
1				
2				
3				
4				
5				
6				
7				
8				
9				
10				
11				
12				
TOTAL				

* * * *

CHAPTER NINE
THE PRICE OF POWER

Introduction

A coal burning power plant located 3 kilometers from a small lake known as Lonesome Lake is put into operation to provide a small village with electricity. Considerable amounts of sulfur (S) are locked up in coal and when coal is burned, the sulfur is released into the air in the form of sulfur dioxide (SO_2) at which stage the SO_2 acquires an additional oxygen molecule to become sulfur trioxide (SO_3). The precipitation absorbs SO_3 and produces sulfuric acid (H_2SO_4), which is deposited along with the precipitation onto the ground and into streams. The presence of excessive amounts of sulfur in the atmosphere means more acid falls with the precipitation. When this happens, the precipitation is called *acid rain*. Acid rain can not only harm vegetation but when rivers and lakes become too acid, fish and other aquatic life die out and drinking water obtained from these sources becomes foul.

This chapter examines the effect of sulfur emission on the water of Lonesome Lake. In order to do this, we look at the annual emission of sulfur from the power plant together with the annual precipitation. The chemical reactions of the sulfur, oxygen and water create hydrogen ions; the acidity of a liquid is measured numerically by a pH scale which is a measure of concentration of hydrogen ions (a formal definition is given below). The greater the amount of sulfur, the greater the number of hydrogen ions; this makes the liquid more acid. A low pH value indicates acid liquid, whereas a higher pH value shows alkaline, or less acid, liquid. A pH value of 7 means the water has neutral acidity. The goal of this chapter is to determine the pH of the water in the lake and the effect this might have on its fish population.

9.1   Sulfur dioxide emission

Not all the sulfur emitted from the power plant ends up over the watershed of the Nizhoni River which feeds the lake. The first step in determining the effect of sulfur emission on Lonesome Lake is to determine the percentage of sulfur which actually does wind up over its watershed. The dispersion of the SO_2 is dependent upon weather conditions, emission rate, and the effective height of the smokestack. We assume a constant emission rate and a fixed weather pattern with the lake situated downwind from the power plant. The *effective height* of the smokestack is the distance from the ground to the horizontal centerline of the smoke plume. (See Figure 9.1.)

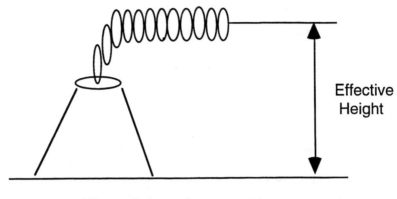

Figure 9.1

   In order to determine the height a procedure which uses properties of right triangles can be used. If one stands at a known horizontal distance from a point directly underneath the plume and measures the angle α between the horizontal and a line to the centerline of the plume this angle is called the *angle of elevation*. A right triangle ABC is formed as shown in Figure 9.2.

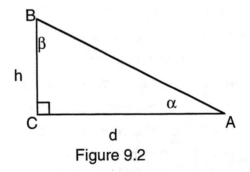

Figure 9.2

The *tangent* of the angle α of elevation (abbreviated tanα) is the length of the side opposite (effective height) divided by the length of the side adjacent (the known distance). If α is the measure of the angle, h is the length of the side opposite, and d is the known distance then

$$\tan \alpha = \frac{h}{d}.$$

The tangent of any angle (other than a right angle) can be determined on the calculator or computer and since d is known this equation can be used to solve for the effective height.

Group Work

If the angle of elevation of the centerline of the smoke plume is 50.2° from a point 100 meters from the base of the smokestack, determine the effective height of the plume (round to the nearest 10 meters).

* * * *

Group Work

The annual emission of sulfur dioxide from the plant is $10^{7.81}$ grams. The emission rate of SO_2, measured in grams per second, is assumed to be constant. Determine the emission rate, and denote it by Q. Use these conversion factors and round your answer to 4 decimal places:

1 year = 365 days;

1 day = 24 hours;

1 hour = 3600 seconds.

* * * *

The ground level concentration of the sulfur dioxide varies with the distance downwind from the stack. The steps below lead you to a function which estimates this concentration at a specified distance downwind. This function is used to approximate the percentage of sulfur emissions which lies over the watershed. As might be expected, the ground-level SO_2 concentration increases from the point of emission to its maximum, then drops off rather rapidly and almost levels out as it approaches 0 at a distance. (See Figure 9.3.)

Figure 9.3

Even though environmental engineers use a more sophisticated function called a Gaussian distribution function to measure concentration, such a pattern can be approximated by a rational function with a denominator of degree larger than that of its numerator. In this case, a function of the form

$$C(x) = \frac{Q}{u}\left[\frac{Ax}{x^3 + B}\right]$$

will give a reasonable approximation. Here,

$C(x)$ = ground-level concentration of sulfur dioxide in g/m^3 at a point x km
      downwind from the stack,

Q = emission rate of SO_2 in g/sec from the top of the stack,

u = wind speed in m/sec,

A and B are constants which depend on the effective height of the smokestack.

For fixed values of Q and u, the effective height of the smokestack determines both the point x_{max} km downwind where the maximum ground-level concentration occurs and the maximum concentration C_{max}. We use data taken from curves developed by Turner (see Masters, [2]) to determine the point (x_{max}, C_{max}) for given effective stack height H and then use this information to derive the important function $C(x)$. The data shown in Table 9.1 provide a relationship between effective stack height, the point of maximum ground-level SO_2 concentration, and the maximum concentration. Here,

H = effective height of the smokestack in meters,

x_{max} = distance in kilometers downwind from the smokestack where the maximum ground-level concentration occurs, and

C_{max} = maximum SO_2 concentration in g/m^3 .

H (meters)	x_{max} (km)	C_{max} (g/m^3)
20	0.20	350.0
40	0.42	90.0
60	0.70	40.0
80	0.90	24.0
100	1.20	15.0
120	1.40	12.0
140	1.70	7.5
160	2.00	5.5
180	2.30	4.1
200	2.50	3.5
220	2.80	3.0
240	3.00	2.5
260	3.30	2.1
280	3.60	1.7
300	4.00	1.5

Table 9.1

IMPORTANT: The data provided in this table are for Q = 1 g/sec. and u = 1 m/sec.; these will be used to first derive a function

$$C_1(x) = \frac{Ax}{x^3 + B}$$

which gives the ground level SO_2 concentration when Q = 1 and u = 1. Then we will only have to multiply or divide (respectively) this function by given values for Q or u to obtain other concentration functions C(x) for different emission rates and wind speeds.

For a particular stack height, this table can be used to determine the point x_{max} kilometers downwind where the maximum ground-level concentration of SO_2 occurs and the SO_2 concentration C_{max} at that point. Using these, it is possible to determine the constants A and B for the concentration function,

$$C_1(x) = \frac{Ax}{x^3 + B}$$

when Q = 1 g/sec. and u = 1 m/sec.

Group Work

Follow the steps below to determine the ground level concentration function C(x).

1.   Determine the values of x_{max} and C_{max} for the effective stack height determined.

2.   Any function of the form $f(x) = \dfrac{Ax}{x^3 + B}$ (where A and B are postive) has a graph of the shape shown above. Using methods of calculus, it can be shown that the maximum point on the graph occurs when $x = \left(\dfrac{B}{2}\right)^{\frac{1}{3}}$ Therefore, if we know when the maximum occurs, i.e., if we know x_{max}, then we can determine B; if we also know the maximum value, i.e., $f(x_{max})$, then we can determine A. Use this information and your answer to #1 to determine the constants A and B. and hence the concentration function $C_1(x)$ when Q = 1 g/sec. and u = 1 m/sec. Round your answer to 3 decimal places.

3.   Use your answer from the previous group work and a wind speed of u = 1 m/sec to determine the function
$$C(x) = \frac{Q}{u}\left[\frac{Ax}{x^3 + B}\right].$$
This will be the concentration function we will use throughout the remainder of this module.

4.   Graph the function C(x) and determine its maximum value. (Use the calculator.)

5.   What is the maximum ground level concentration and where does it occur?

* * * *

Now we will determine the percentage of ground level SO_2 which actually is distributed over the watershed. At distances over 15 kilometers from the plant, SO_2 concentrations are insignificant, so we will approximate the percentage of

SO_2 distributed over the watershed relative to the amount which is distributed from the point of emission (smokestack) to the point 15 kilometers downwind.

Group Work

1. The units of measure for $C(x)$ are grams per cubic meter of SO_2. For example if $C(3) = 13.3$ g/m^3, this means that one cubic meter of air on the ground at 3 km from the smokestack contains 13.3 grams of sulfur dioxide. Now imagine lots of cubic meters of air lined up on the ground from the smokestack to 15 km downwind. This would form a "box" of air one meter wide, one meter high, and 15 km long. (See figure.)

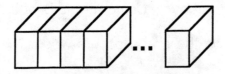

A) Approximate the amount of SO_2 in this "box" of air. Assume that the SO_2 concentration is $C(x)$ for each kilometer from $x - 1$ to x. (Note: using the above example, each cubic meter of air in this box from 2 km to 3 km contains 13.3 grams of SO_2. Since 1 km = 10^3 m, the box from 2 km to 3 km contains $13.3 \times 10^3 = 13300$ grams of SO_2.)

B) See Figure 9.4 to understand the relative locations of the power plant, lake, and watershed. Lonesome Lake is 3 km downwind from the power plant with the watershed extending 3 km further downwind from the lake. Approximate the total amount of SO_2 in the "box" of air through the watershed.

2. Use your answers from #1 to approximate the percentage of SO_2 which lies over the watershed. (Although this approximation only takes into account the sulfur dioxide directly downwind from the power plant we can assume that the same ratio holds over the entire region.)

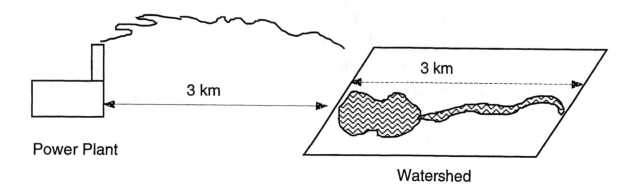

Power Plant

3 km

3 km

Watershed

Figure 9.4

* * * *

9.2. Acid rain

You are now ready to begin work on the determination of the overall effect of SO_2 emission on Lonesome Lake. The next step is to figure the acidity of precipitation in the region.

The acidity of a liquid is measured by its hydrogen ion concentration, and is denoted by pH. If z denotes the hydrogen ion concentration in moles per liter, then

$$pH = -\log z.$$

The pH varies from 0 to 14; a graph is shown in Figure 9.5.

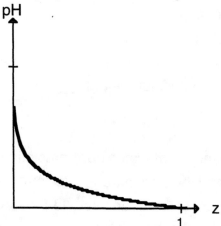

Figure 9.5

Note that as the hydrogen ion concentration z increases, the pH decreases. Since the larger the hydrogen ion concentration the more acidic the liquid, then small values of pH indicate an acid liquid, whereas large pH values indicate alkalinity. Water with pH = 7 is neutral and considered "pure" whereas water with pH smaller than 7 is considered acidic. Thus for pure water, the hydrogen ion concentration is 10^{-7}, and

$$pH = -\log(10^{-7}) = -(-7) = 7.$$

The steps below will lead you to the determination of the pH of precipitation over the watershed. For simplicity, the only sulfur dioxide source we consider in these calculations is the emission from the power plant. We assume:

1)   the total emission of SO_2 from the plant is $10^{7.81}$ grams per year;

2)   the total amount of water from the precipitation over the watershed is $10^{9.45}$ liters per year; and

3)   one-fourth of the air born sulfur is deposited in precipitation.

Group Work

1.   The total annual emission of SO_2 from the plant is $10^{7.81}$ grams per year. Our primary concern is the amount of sulfur emitted since sulfur is the component which forms the sulfuric acid. Sulfur contributes approximately one-half the weight of sulfur dioxide; the atomic weight of one sulfur molecule is 32.066 awu (atomic weight units) whereas one oxygen molecule has atomic weight 16 awu. Determine the annual emission of sulfur from the plant (measured in grams).

2.   Since concentration is measured in moles per liter, the sulfur must be measured in *moles*. Determine the number of moles of sulfur in the annual SO_2 emission from the power plant. You need this information. One mole of substance is equal to its molecular weight in grams; so 32.066 grams of sulfur equal one mole.

3.   Recall the total amount of water from precipitation over the watershed is $10^{9.45}$ liters per year and approximately one-fourth of the airborne sulfur is deposited in precipitation. Use your answers to #2 above and #2 in the previous group work to determine the concentration of sulfur (in moles per liter) in precipitation over the watershed.

4.   For every molecule of sulfuric acid, two hydrogen ions are produced; i.e. for each molecule of H_2SO_4 (which contains one sulfur molecule), two hydrogen ions (H^+) result. Determine the concentration (in moles per liter) of hydrogen ions in precipitation over the watershed.

5.   What is the pH of the precipitation?

* * * *

9.3.   Acidity of the water in Lonesome Lake

From this point, two things happen which affect the pH of the precipitation before it reaches Lonesome Lake. First, some of the precipitation immediately evaporates and then a significant evaporation of the surface water in the lake occurs; both of these events further concentrate the sulfuric acid. Overall, a total of one-forth of the total precipitation evaporates. In this process the amount of sulfuric acid remains constant so that there is an even higher concentration in the lake water. Thus the concentration of H^+ ions is higher and the pH is lower. In the steps below, you will determine the acidity of the water in the lake; we assume that no other chemical reactions take place which will further affect the pH of the water.

Group Work

1. Determine the hydrogen ion concentration of the water after the evaporation.

2. What is the pH of the water in Lonesome Lake?

* * * *

9.4.   Fish

It has been established that acid water is harmful to fish life. In fact, a study of declining fish populations in acidified freshwater lakes in Norway was conducted by Leivestad, Hendrey, Muniz, andSnekvik in the 1970's. The findings are quite conclusive; fish simply do not survive in waters which have low pH. The absence of fish life in freshwater is not only ecologically unsound, but also affects commercial and recreational activities.

Group Work

The bar graph shown in Figure 9.6 indicates some of the results of the Leivestad *et al* study. Examine this graph and determine the probability that Lonesome Lake has any fish. Then examine the variables involved in this module to see what can be done to improve this probability. The variables are:

i) emission rate Q (this could probably be reduced some, but if you want electricity, not too much);

ii) wind speed (of course you have no control over this, but you can see how increased wind speed affects dispersion of sulfur dioxide);

iii) stack height H (this is one of the main factors used in controlling SO_2 dispersion from coal burning power plants).

Write a report explaining your plan to reduce the acidity of Lonesome Lake so that the fish have a chance.

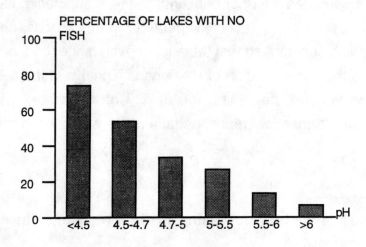

Figure 9.6

* * * *

CHAPTER TEN
ARSENIC

Introduction

Development, agriculture, mining and industrial activity not only increase the amount of water used, these activities also effect the quality of the water. Mining is particularly hard on surrounding water sources. In this chapter we examine the effect of a gold mine near the Nizhoni River. Many substances are used in the processing of gold, and we will examine the effect of one of these, arsenic, on the water supply of River City.

A gold mine built near the river upstream from River City produces over 6000 metric tons of waste for each kilogram of gold produced, and a portion of that waste seeps into the river and downstream into the water supply system. The mine discharges 15 pounds of arsenic each day in the waste that enters the river and mixes with the water before flowing into the reservoir. Although a small amount of arsenic has no noticeable effect, if the level gets to high the water quality suffers. Since the reservoir is used for drinking water and since the residents eat fish from the river, the people of the town are concerned about the degradation of the water supply. Government regulations limit the concentration of arsenic in drinking water to .05 milligrams per liter of water. The residents of River City begin monitoring the quality of water, testing for arsenic to ensure that the level remains within the government guidelines.

10.1 Arsenic Concentration

The concentration is measured in weight per volume. The regulations use milligrams per liter but we can use any convenient units of weight and volume. In Chapter 8 you computed the volume of water in the reservoir in million cubic feet, and since the amount of arsenic entering the river is given in pounds, we will first measure the concentration in pounds per cubic feet. The concentration varies with time; arsenic enters the river, mixes with the water, flows through the reservoir and flows out. The concentration depends not only on the amount of arsenic entering the reservoir and the volume of water in the reservoir but also on the amount of time it takes the water together with the polluting substance to flow out of the reservoir.

Our model assumes that there is no mixing with the lake sediment. Some of the arsenic entering the lake may precipitate out depending on the pH of the lake. At other times some of the arsenic in the sediment may mix back into the lake water. Here again this re-mixing is pH dependent. The actual mathematical model for determining arsenic pollution or any other pollutant is much more complex.

10.1.1   Average Flow, Average Volume

We start with a simplified case, where the volume and flow are constant. Although the actual volume of water in the reservoir and flow of water through the reservoir varies over time, we will begin with a simplified study. Assume that the volume of water in the reservoir is constant, 37.5 million cubic feet, and the flow of water into and out of the reservoir is constant 180 million cubic feet per month.   (These are "averages" for the volume and flow from Chapter 8.) Assume also that arsenic enters the reservoir at a constant rate, 15 pounds per day. Let c(t) denote the concentration (pounds per 10^6 cubic feet) and m(t) the amount (pounds) of arsenic in the reservoir at time t. We will answer a series of questions to determine whether the water quality meets the EPA standards.

Question 1   How much water flows into (and out of) the reservoir each day?

The flow of water is given per month so we convert this to daily flow,

$$180.0 \times 10^6 \text{ cubic feet per month} =$$
$$180 \div \frac{365}{12} \times 10^6 \text{ cubic feet per day} =$$
$$6 \times 10^6 \text{ cubic feet per day (rounded)}.$$

Question 2   How much arsenic flows in each day?

This is easy, fifteen pounds per day.

Question 3   What portion of the arsenic in the reservoir flows out each day?

The amount flowing out depends on the amount present, but a fixed portion flows out each day. The amount of arsenic flowing out each day is the concentration of arsenic in water times the amount of water flowing out (6×10^6 cubic feet). Since the concentration is $\dfrac{m(t)}{37.5}$ pounds per 10^6 cubic feet, arsenic flows out at a rate of $6 \times \dfrac{m(t)}{37.5} = .16 \times m(t)$ pounds per *day*. That is, 16% of the arsenic present flows out each day and 84% remains.

Question 4 How much arsenic is there in the reservoir at any time t? Determine a formula for the function m(t) which describes this.

To determine a formula, first notice that each day 15 pounds of arsenic is added to the 84% of "old" arsenic that remains, $m(t+1) = 15 + .84m(t)$.

$m(1) = 15 + 84(0) = 15$

$m(2) = 15 + 84(15) = 15(1 + .84)$

$m(3) = 15 + .84\{15(1 + .84)) = 15(1 + .84 + .84^2)$

. . .

At the end of t days the amount of arsenic in the lake is given by the formula

$m(t) = 15(1 + .84 + .84^2 + ... + .84^{t-1})$;

notice that this is a geometric series and recall the formula for the sum (see section 10.3 for this derivation and more on geometric series),

$$a\left(1 + r + r^2 + ... + r^{n-1}\right) = a \cdot \frac{1 - r^n}{1 - r}.$$

Therefore after t days the amount of arsenic present is

$$15\left(1 + .84 + .84^2 + ... + .84^{n-1}\right) = 15 \cdot \frac{1 - .84^n}{1 - .84} = 93.75(1 - .84^n),$$

i. e.,

$m(t) = 93.75(1 - .84^t)$.

Question 5 What is the concentration on day t? Determine a formula for this.

The concentration is simply the amount divided by volume,

$$c(t) = \frac{93.75(1 - .84^n)}{37.5} = 2.5(1 - .84^n) \text{ pounds per } 10^6 \text{ cubic feet}.$$

Question 6   What happens to concentration of arsenic in the water over a long period of time?   Use the graph of c(t) and interpret its asymptote (see Figure 10.1).

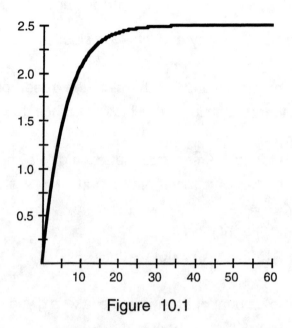

Figure 10.1

The graph appears to be leveling off, and this apparent trend can be verified if we take a closer look at the function $c(t) = 2.5(1-.84^n)$. This is an exponential function with base less than 1. The graph has the horizontal asymptote $y = 2.5$ and the range consists of all $y < 2.5$ and hence the concentration is always less than 2.5 pounds per million cubic feet.

Question 7   What is the highest level (in milligrams per liter) that the concentration reaches?

A conversion is required,

$$\frac{1 \text{ pound}}{10^6 \text{ cubic feet}} = \frac{.016 \text{ milligram}}{\text{liter}}.$$

Then the highest level of concentration reached is 2.5 x .016 = .040 mg per liter, so the concentration stabilizes at a rate within the federal guidelines.

* * * *

10.1.2   Change in Concentration.

In the example above we determined the concentration of arsenic using the average volume and average flow for the reservoir. In the group work that follows examine what would happen during a low volume, low flow month.

Group Work

Repeat the study above, now assuming that the volume and flow are close to those which you determined in chapter 8 for March; the actual volume and flow you got depended on round-off but the answers should be close to a volume of 13.5 million cubic feet and flow of 28.4 million cubic feet per month. Assume:

the volume of water in the reservoir is constant, 13.5 million cubic feet;

the flow of water is constant, 28.4 million cubic feet per month;

arsenic enters the reservoir at a constant rate, 15 pounds per day.

Again, let c(t) denote the concentration (pounds per 10^6 cubic feet) and m(t) the amount (pounds) of arsenic in the reservoir at time t and follow the steps below.

1. How much water flows into (and out of) the reservoir each day?
2. How much arsenic flows in each day?
3. What portion of the arsenic in the reservoir flows out each day?
4. Determine a formula for the function m(t) which describes the amount of arsenic in the reservoir at time t.
5. Determine a formula for the concentration of arsenic on day t. Graph the function and determine any asymptotes.
6. What is the highest level (in milligrams per liter) that the concentration reaches?
7. What happens in the long run? In particular, does the water ever fail to meet the federal guidelines of .05 milligrams per liter? If so, when?

★ ★ ★ ★

10.1.3   A General Formula

In the work above we assumed that there was no arsenic present initially and that the volume of the reservoir (V), the daily rate of water flow (F) and the rate of arsenic flowing into the water (A) were all constant. We can now easily derive a general formula for the concentration of arsenic in the water. To do this, we again answer questions 1 - 4 above.

1.   F is the daily flow.

2.   A pounds of arsenic flow in each day.

3.   The portion which flows out each day is F/V and r = 1 - F/V is the portion remaining.

4.   m(t + 1) = A + r m(t); the pattern is the same as before.

That is,

$$m(1) = A, m(2) = A + rA, m(3) = A + rA + r^2A,$$

etc., and as before this gives us

$$m(t) = A(1+r+r^2+...+r^{t-1}) = A \times \frac{1-r^t}{1-r} = \frac{A}{1-r}(1-r^t).$$

Note that

$$r = 1 - \frac{F}{V}$$

and so

$$\frac{A}{1-r} = \frac{A}{F/V} = A \times \frac{V}{F}$$

and

$$m(t) = A \times \frac{V}{F} \times (1-r^t).$$

Note that this interpretation makes sense provided that F < V; if the flow is greater than the volume, change the unit of time. For example, if the flow is 4.8 million cubic feet per day and the volume is only 4 million cubic feet, then change to hours. Then the flow is .2 million cubic feet per hour, the volume is still 4 million cubic feet, .2/4 = .05 is the portion flowing out and 95% of the water remains each hour.

If there is already arsenic present in the water, say that the amount present is p pounds, then a slight modification is necessary in the last step.

$$m(0) = p, m(1) = A + rp, m(2) = A + rA + r^2p, m(3) = A + rA + r^2A + r^3p, \text{etc.,}$$

and this gives us

$$m(t) = A(1+r+r^2+...+r^{t-1})+r^tp.$$

The term r^tp has been added and so

$$m(t) = A \times \frac{V}{F} \times (1-r^t)+r^tp.$$

If t is very large, this new term is negligible.

* * * *

10.2 Arsenic And Risk

The risk from toxic substances in the water depends on many factors: the concentration of the substance; the extent of the exposure; the use of the water; etc. Water that might be safe for boating could be unsafe for drinking water; water acceptable for swimming might contain fish with high levels of toxins in their tissues. The Environmental Protection Agency provides information to help determine the risk. The Table 10.1 below provides information about average human size and intake of water and fish.

	Weight (kilograms)	Daily water intake (liters)	Daily fish consumption (grams)
Adult	70	2	6.5

Table 10.1

Toxic substances in water can be dangerous to humans either directly through the water they drink or indirectly if they accumulate in the tissue of fish which are then eaten by humans. The measures defined below help determine the risk of toxic substances in water.

The measure of the tendency for a substance to accumulate in the tissues of fish is called the *bioconcentration* factor and is measured in liters per kilogram. The concentration of the substance in fish is the concentration in water (mg per liter) times the *bioaccumulation* factor (liters per kilogram). The *average daily dose* measures the milligrams of substance per kilogram of body weight consumed per day (mg/kg/day). The *potency factor* is the risk produced by a lifetime daily dose of 1 mg/kg/day and provides a measure for determining the *lifetime risk,*. This is the probability of getting cancer over an average (70 year) life span,

Lifetime Risk = (average daily dose) x (potency factor).

Table 10.2 provides:

(mg

the maximum acceptable concentration level (MCL) for arsenic
per liter of water);

the bioconcentration factor (liters per kilogram; and

the potency factor.

	MCL (mg per liter)	Bioconcentration (liters per kg)	Potency
Arsenic	0.05	44	1.75

Table 10.2

Use this information to complete the following examination of the long-term effects of arsenic in the drinking water of the residents of River City.

Group Work

To study the long-term effects of arsenic on the residents of River City we will assume that the volume of water in the reservoir is constant, the flow is constant, the influx of arsenic is constant, and therefore we can the information above and assume that the concentration has stabilized at .040 mg per liter.

1.   Fill in the table below to determine risks from arsenic from drinking water and eating fish out of the reservoir.

Source	Concentration	Daily intake mg/day	Ave. daily dose mg/kg/day	Risk
Water				
Fish				

2.   Use the risks determined in the table above to predict the number of cancer cases per thousand residents of River City.

3.   Unfortunately for the inhabitants of River City, the reservoir volume is not constant. During months when volume is low and residence time is high the concentration is greatly increased. To see the significance of this, determine the concentration at the end of the month of March (as determined in the group

work above) and assume that the initial amount of arsenic present is zero. What would be the lifetime risks if the concentration remained at this level?

4.   Determine the maximum daily inflow of arsenic so that the March concentration does not exceed the acceptable level of .05 mg/L.

$$* \quad * \quad * \quad *$$

10.3 Geometric Series

Earlier in this chapter we used properties of *finite geometric series* to study the accumulation of arsenic in the water. In this section we will take a closer look at the ideas we used.

A *finite geometric progression* is a sequence of numbers, each of which is obtained by multiplying the previous number by a fixed constant. The general form for a geometric progressions is

$$a, ar, ar^2, ..., ar^n,$$

where a can be any number, r is the fixed constant, and n is any positive integer.

A *finite geometric series* or simply *geometric series* is

$$a + ar + ar^2 + ... + ar^n,$$

i.e., the sum of a finite geometric progression. The sum

$$S = a + ar + ar^2 + ... + ar^n$$

of a finite geometric progression is given by the formula

$$S = \frac{a(r^{n+1} - 1)}{r - 1}, (r \neq 1).$$

After a few examples of using this formula we will show its derivation.

Example 10.1

The series 5 + 10 + 20 + 40 + 80 + 160 is a geometric series. Determine its sum.

Solution

First determine

$$a = 5,$$

$$r = 2,$$

$$n = 5.$$

Substitute into the formula to get

$$S = \frac{5(2^6 - 1)}{2 - 1} = 315.$$

* * * *

Example 10.2

Determine the sum of the geometric series.

Solution

First determine

a = 5,

r = 2,

n = 5.

Substitute into the formula to get

$$S = \frac{5(2^6 - 1)}{2 - 1} = 315.$$

* * * *

The formula which gives the sum of a finite geometric series is not hard to derive. Multiply both sides of the series

$$S = a + ar + ar^2 + ... + ar^n$$

by the common ratio r,

$$rS = ar + ar^2 + ar^3 + ... + ar^{n+1};$$

subtract the second equation from the first, to get

$$rS - S = (ar + ar^2 + ar^3 + ... + ar^{n+1}) - (a + ar + ar^2 + ... + ar^n);$$

simplification gives

$$(r - 1)S = a(r^{n+1} - 1).$$

Finally, solve for S to get

$$S = \frac{a(r^{n+1} - 1)}{r - 1}.$$

* * * *

Exercises

In exercises 1-5, determine the numbers a, r, and n, and the sum of each geometric series.

1. $1 + 3 + 9 + 27 + 81 + 243 + 729$

2. $2 + .2 + .02 + .002 + .0002 + .00002$

3. $100 + 100(1.05) + 100(1.05)^2 + ... + 100(1.05)^{10}$

4. $8 + 4 + 2 + 1 + \dfrac{1}{2} + \dfrac{1}{4} + \dfrac{1}{8} + \dfrac{1}{16}$

5. $10 + 11 + 12.1 + 13.31 + 14.641$

6.   The formula for the sum of a geometric series is not valid when r = 1. What is the sum of the series when r = 1?

* * * *

THIS IS THE END, BEAUTIFUL FRIEND

APPENDIX A
PROOFS

A.1 The Law of Cosines

Law of Cosines : in the triangle shown (Figure A.1),

$$a^2 = b^2 + c^2 - 2bc\cos\alpha .$$

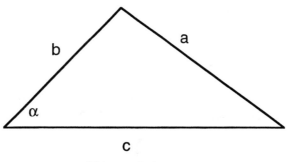

Figure A.1

The derivation of the formula for the law of cosines is not difficult if you look carefully at an appropriate picture and think hard about the definitions of the trigonometric functions. First place the triangle ABC in a coordinate system with vertex A at the origin and side AB along the positive x-axis. Two possible pictures, depending on whether α is acute or obtuse, are shown in the Figure A.2.

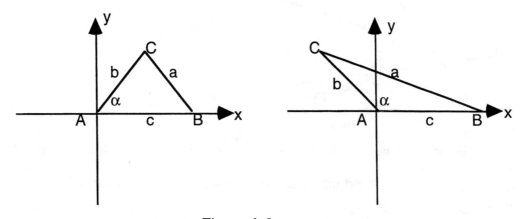

Figure A.2

The coordinates of the points A and B are easy to determine,

A = (0,0) and B = (c,0).

If C = (x,y), then $\sin\alpha = \dfrac{y}{b}$ and $\cos\alpha = \dfrac{x}{b}$ and so $y = b\sin\alpha, x = b\cos\alpha,$ so we can rewrite C = (bcosα, bsinα). Now we consider the length of the side CB. This length is *a*, but we can also use the distance formula

$$\sqrt{(b\cos\alpha - c)^2 + (b\sin\alpha - 0)^2} = a,$$

now square both sides and simplify,

$$(b\cos\alpha - c)^2 + (b\sin\alpha - 0)^2 = a^2$$

$$b^2\cos^2\alpha - 2bc\cos\alpha + c^2 + b^2\sin^2\alpha = a^2$$

$$b^2(\cos^2\alpha + \sin^2\alpha) - 2bc\cos\alpha + c^2 = a^2$$

$$b^2(1) - 2bc\cos\alpha + c^2 = a^2$$

$$b^2 + c^2 - 2bc\cos\alpha = a^2.$$

The other two forms of the law of cosines follow in a similar manner.

* * * *

A.2   The Law of Sines

Law of Sines: In the triangle ABC (Figure A.3) ,

$$\frac{\sin\alpha}{a} = \frac{\sin\beta}{b} = \frac{\sin\gamma}{c}.$$

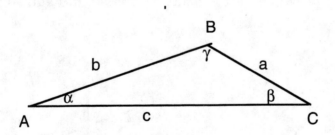

Figure A.3

We will prove the equality $\dfrac{\sin\alpha}{a} = \dfrac{\sin\beta}{b}$.   Figure A.4 first shows the triangle with angle α in standard position; the second shows the triangle with angle β in standard position.

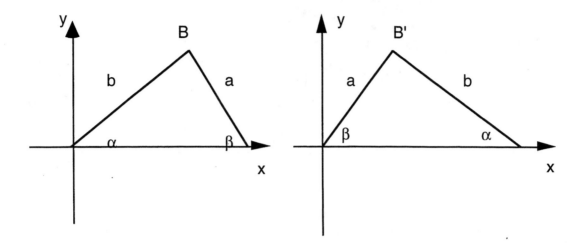

Figure A.4

The points B and B' can be written

      B = (bcosα, bsinα) and B' = (acosβ, asinβ),

respectively.   Since the two pictures show the same triangle in different positions, then the points B and B' have the same height and hence the same y-coordinate it follows that

      bsinα = asinβ,

or

$$\frac{\sin \alpha}{a} = \frac{\sin \beta}{b}.$$

The other equation

$$\frac{\sin \alpha}{a} = \frac{\sin \gamma}{c}$$

follows in a similar manner.

* * * *

A.3   Sum and Difference Identities

(1) $\cos(\alpha - \beta) = \cos \alpha \cos \beta + \sin \alpha \sin \beta.$

(2) $\cos(\alpha + \beta) = \cos \alpha \cos \beta - \sin \alpha \sin \beta.$

(3) $\sin(\alpha + \beta) = \sin \alpha \cos \beta + \cos \alpha \sin \beta.$

(4) $\sin(\alpha - \beta) = \sin \alpha \cos \beta - \cos \alpha \sin \beta.$

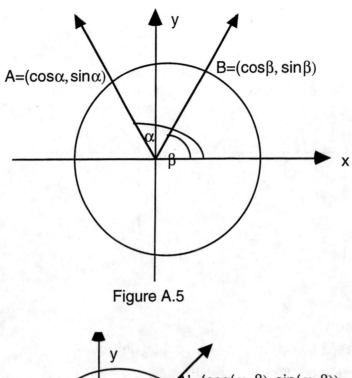

Figure A.5

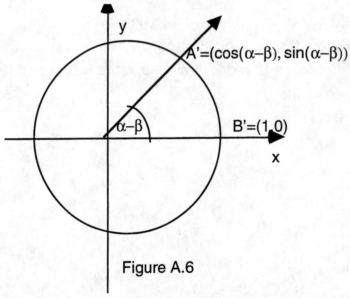

Figure A.6

Proof of 1. Refer to Figures A.5 and A.6. This proof is very geometric and uses the distance formula. We consider the special case where the two angles α and β are in standard position with $\alpha > \beta > 0$. (The general case can be proved in a manner similar to the one given here.) We will first compute the distance between the points A and B. Then consider the angle $\alpha - \beta$ in standard position and compute the distance between the points A' and B'. These two distances are clearly the same; by equating the squares of these distances and simplifying, we obtain identity #1.

First refer to Figure A,5. The points A and B of intersection of the terminal sides of angles α and β with the unit circle can be written

$A = (\cos\alpha, \sin\alpha)$ and $B = (\cos\beta, \sin\beta)$.

Then the square of the distance between A and B is

$(\cos\alpha - \cos\beta)^2 + (\sin\alpha - \sin\beta)^2$.

Now expand and simplify this expression,

$\cos^2\alpha - 2\cos\alpha\cos\beta + \cos^2\beta + \sin^2\alpha - 2\sin\alpha\sin\beta + \sin^2\beta$

$= (\cos^2\alpha + \sin^2\alpha) + (\cos^2\beta + \sin^2\beta) - (2\cos\alpha\cos\beta + 2\sin\alpha\sin\beta)$

$= 1 + 1 - 2(\cos\alpha\cos\beta + \sin\alpha\sin\beta)$

$= 2 - 2(\cos\alpha\cos\beta + \sin\alpha\sin\beta)$.   (1)

Now consider Figure A.6. The point A' at the intersection of the terminal side of angle $\alpha - \beta$ and the unit circle can be written

$A' = (\cos(\alpha - \beta), \sin(\alpha - \beta))$,

and the point B' on the x-axis and unit circle is

$B' = (1, 0)$.

The square of the distance between A' and B' is

$[\cos(\alpha - \beta) - 1]^2 + [\sin(\alpha - \beta) - 0]^2$.

Then expand and simplify,

$\cos^2(\alpha - \beta) - 2\cos(\alpha - \beta) + 1 + \sin^2(\alpha - \beta)$

$= [\cos^2(\alpha - \beta) + \sin^2(\alpha - \beta)] - 2\cos(\alpha - \beta) + 1$

$= 1 - 2\cos(\alpha - \beta) + 1$

$= 2 - 2\cos(\alpha - \beta)$.   (2)

Now equate expressions (1) and (2) and solve for $\cos(\alpha - \beta)$,

$2 - 2\cos(\alpha - \beta) = 2 - 2(\cos\alpha\cos\beta + \sin\alpha\sin\beta)$,

$\cos(\alpha - \beta) = \cos\alpha\cos\beta + \sin\alpha\sin\beta$,

which is identity #1.

★ ★ ★ ★

Proof of #2. This one is much easier. Use basic identities 5 and 6, write $\alpha + \beta$ as $\alpha - (-\beta)$ and apply the identity for $\cos(\alpha - \beta)$,

$\cos(\alpha + \beta) = \cos[\alpha - (-\beta)] = \cos\alpha\cos(-\beta) + \sin\alpha\sin(-\beta)$

$= \cos\alpha\cos\beta - \sin\alpha\sin\beta$.

★ ★ ★ ★

Proof of #3. This is also easy. Use basic identities #7 and 8,

$$\sin(\alpha + \beta) = \cos[\pi/2 - (\alpha + \beta)] = \cos(\pi/2 - \alpha)\cos\beta + \sin(\pi/2 - \alpha)\sin\beta$$
$$= \sin\alpha\cos\beta + \cos\alpha\sin\beta,$$

which is identity #3.

* * * *

Proof of #4. This is similar to the proof of #3. Use basic identities #5 and #6 and the identity for $\sin(\alpha + \beta)$,

$$\sin(\alpha - \beta) = \sin[\alpha + (-\beta)] = \sin\alpha\cos(-\beta) + \cos\alpha\sin(-\beta)$$
$$= \sin\alpha\cos\beta - \cos\alpha\sin\beta,$$

which is identity #4.

* * * *